AF342652

THE CUNNING SEEDS

Dr. Cecil Sinnamon

First published in 1982 by
Science Teachers Association of Queensland.
Second and revised edition published in January, 1983
by BOOLARONG PUBLICATIONS for
Dr. Cecil Sinnamon,
26 Ryan Road, St. Lucia. Qld. 4067.

National Library of Australia
Cataloguing-in-Publication data

Sinnamon, Cecil, 1904–
    The cunning seeds.

    Originally published: Brisbane: Science Teachers
    Association of Queensland, 1982.
    ISBN 0 908175 54 X.

    1. Micro-organisms. I. Title.
576

Design, reproduction and photo-typesetting
in 11½ on 12½ pt Baskerville by
Press Etching Pty. Ltd., Brisbane.
Printed by The Dominion Press, Hedges & Bell, Melbourne.

THE CUNNING SEEDS

The concept of "seeds" of infectious living matter is not new.

Lucretius, one of the best of Roman poets and thinkers writing "On The Nature of Things" in the first century B.C. spoke of "the seeds of disease", but it was Sir Ronald Ross who first used the phrase "Thy cunning seeds" in the hymn of thanks he wrote in celebration of his discovery in 1898 of the mosquito transmission of malaria.

The monument bearing this inscription, which is shown on the front cover, provides an entrance gate to the Medical School in Calcutta.

ABOUT THE AUTHOR

Dr. Cecil Sinnamon holds the following formal academic qualifications:

Diploma of Pharmacy, Univ. Queensland
Bachelor of Medicine, Univ. Queensland
Bachelor of Surgery, Univ. Queensland
Diploma of Tropical Medicine, Univ. Sydney
Diploma of Tropical Hygiene, Univ. Sydney
Diploma of Microbiology, Univ. Manchester

After returning from active service in 1946, he was appointed Senior Lecturer in Microbiology and Tropical Diseases at the University of Queensland Medical School. In 1951 he was appointed Senior Research Fellow of the Queensland Institute for Medical Research, in which capacity he established the Innisfail Field Station to investigate the unknown fevers of North Queensland.

His work there on leptospirosis resulted in publication of papers on the discovery of Canicola fever, the serological classification of eighty-nine strains of leptospirae, and a new appreciation of the clinical features and the epidemiology of leptospirosis which has been of great value, especially to the sugar industry in Queensland.

"To Nell, my wife, who after fifty-five years of happy companionship, died of cancer".

In memory of her, all profits from the sale of this book will go to the Queensland Cancer Fund.

CONTENTS

ACKNOWLEDGEMENT

I wish to thank the Queensland Science Teachers Association for first publishing parts of this work and recommending it as suitable for use in secondary schools. Mrs. Cecily Gredden, the President of the Association and Mr. Robert McAllister of the Department of Education for their encouraging reviews and helpful suggestions; my daughters Mrs. Bardwell and Mrs. Akers and my grand-daughter Judith Sinnamon for artwork and typing.

INTRODUCTION

THE CUNNING SEEDS

*"To give to airy nothing
A local habitation and a name."*

Midsummer Nights Dream

This is a book about the invisible.

Not wandering ghosts and shades which have no substance and no use or value except to frighten the credulous, but about living active microscopic beings on whose remarkable activities, though invisible, the very existence of the visible world depends.

Not being conspicuous, they are little known, and, when thought of at all they are seen, not as the indispensable custodians of the world of living matter, but as dangerous carriers of death, delighting in epidemic slaughter, something foul to be avoided or destroyed; something to be feared and hated.

Hideous gruesome germs!

Yet these imperceptible organisms are Man's constant servants who provide his food, his wine and his clothing. They have changed the sterile surface of the planet into fertile soil and the nitrifying bacteria continue to convert hundreds of millions of tons of atmospheric nitrogen each year into fertilising nitrates for the farmers' crops.

The baker needs the help of the active yeasts, and the dairyman's cheese and butter owe their richness and flavour to the obliging streptococci.

The products of brewer and vintner depend on the micro-biologist's careful control. Not only does the microscopic life of the Earth provide Man with food — it also helps to digest it. In the intestines of herbivorous animals, it changes the indigestible plant polysaccharide — the cellulose — into digestible compounds; in the alimentary tract of man himself, it manufactures vitamins for his health; even in the gut of the lowly termite, filled with the dense products of its own small wood-chip industry, it changes the hard wood into termite food.

The environmentalist and the antipolluntionist would be buried in deserts of rotting wastes without these tireless unpaid garbage men who purify and re-cycle the pouring torrents of sewage from multimillioned cities, and return the lifeless bodies of all creatures, great and small, to the earth whence they came. Even the fallen forest giant must moulder and disintegrate under their constant attacks. And the proud body of Man too, when he dies, is dismantled and returned element by element to the same earth by these invisible servants who accompany him from his first moments till his last.

Man uses microscopic organisms in processes too involved for his own technicians. In his ceaseless search for agents of destruction and war, his most deadly poisons are several thousand times less potent than the subtle toxins produced by the bacteria which lie unseen in the garden soil where he grows his fragrant roses.

The bacterial rods of tetanus and botulism in their normal soil habitat help to cleanse the Earth of organic wreckage and corruption and return it to fertilise the fields again, but in wounded flesh may cause violent convulsive death, or can engender fantastically lethal poisons. Aggressive Man now hopes to use them too as a new weapon against his enemies but long ago the unsophisticated savage learned the fatal power of such toxins on the tip of spear or arrow.

What then do we know about this essential hidden part of the living world, so helpful in its normal moods, so deadly when its intricate bio-chemistry is altered by some environmental accident? Systematic study began when his magnifying lenses showed their presence to an observant Dutchman and has progressed through the development of the simple microscope, the compound and the electron microscopes and a multitude of techniques acquired by patient laboratory study which are now leading us on, it is hoped, to a knowledge of life's

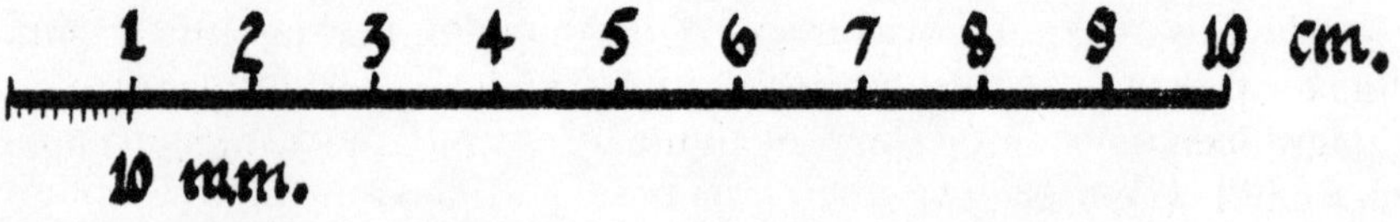

Microorganisms are measured in microns and millimicrons.
A millimetre is the one thousandth part of a metre.
A micron is the one thousandth part of a millimetre.
A millimicron is the one thousandth part of a micron.

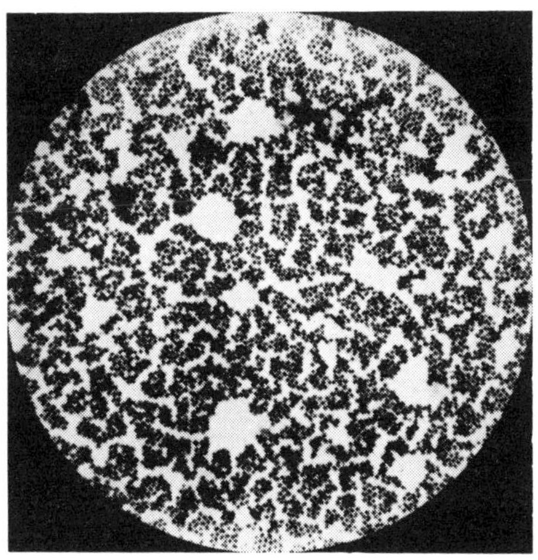

This is a microphotograph of Staphylococcus Aureus organisms — the well-known "Golden Staph".

These organisms are one micron in diameter and in this photograph they are magnified x 900 so that one micron is one nine hundredth of their diameter and one millimicron is one nine-hundred thousandth of the diameter of each. It is impossible to visualise such small measurements.

origins. They are measured on a scale which uses microns — the thousandth part of a millimetre — and millimicrons — the thousandth part of a micron and even by the Angstrom unit which is used to measure the wavelength of light and is ten to the power of minus ten or one ten thousand millionth of a metre.

They are grown on a variety of curious media, classified by criteria which include their shape and size, spores and mobility, the dyes which they absorb, their chemical affinities, their enzyme systems and metabolism, and, of course, because of Man's preoccupation with his own bodily safety and that of his animals, by their pathological effects.

They may be in smallness down to the very ultimate of possible living matter, yet capable of chemical and other activities which suggest intellectual powers and memory.

Though most of us are not so fortunate as to see these interesting beings for ourselves, the microbiologist with the aid of optical systems which allow them to be seen and photographed, and equipment which makes possible their measurement and analysis, can display to us the magnificence of their minute architecture and the wonders of their incredible physiology.

Optical examination of such organisms by compound light microscopes, whose limit of magnification is between 1,000 and 2,000 diameters gives much information on the gross aspects of the larger organisms but cannot show their cellular structure or demonstrate those which are below the limits of such optical magnification. The development of the electron microscope in which a stream of electrons replaces the beam of light and interposed objects cast a shadow which can be photographed, now gives the microbiologist a possible resolution of up to 100,000 diameters and with it he is able to probe down to the smallest masses which, it is thought, could be consistent with the activities of living matter and independent existence. Somewhere in this indistinct borderland, there exist viruses which are mere molecules of only a few genes, parasites of organisms which are themselves microscopic. Somewhere here the living and the non-living meet. Somewhere here viruses, artificially changed into crystals of nucleic acid, may be changed back into living proteins. It is as though living matter having divided and diminished itself until it reached the incredible minuteness of only a few genes, and is about to lose its identity in the lifelessness of chemical substitutes, draws itself back from the dangerous brink and becomes living matter again.

If Man ever learns how life has evolved from the non-living as he thinks it has, he will owe yet another debt to the humble microbe, for it is on their single-celled and strangely efficient bodies that his researching microscopes are now mostly focussed. At present, he is very far from finding their elusive origins.

The microbes hold that tantalising secret and have obstinately refused to give it up.

OUR EARTH-BORN COMPANIONS AND FELLOW MORTALS

Wee, sleekit, cow'rin', tim'rous beastie
O what a panic's in thy breastie!
Thou need na start awa sae hasty,
Wi' bickering brattle!
I wad be laith to rin an' chase thee,
Wi' murd'ring prattle!

I'm truly sorry Man's dominion
Has broken Nature's social union,
And justifies that ill opinion
That makes thee startle
At me, thy poor Earth-born companion
And fellow-mortal.

To a mouse, on turning her up in her
nest with a plough.
November 1785
Robert Burns

To the untrained mind mystery appeals more than fact. It was natural then that before scientific thought developed, Man turned to the supernatural to explain the unknown. So long as he knew only the visible world around him he saw himself as the divinely appointed Master of all living things, but biology, including the sciences of microbiology, physiology, biochemistry and genetics has taught him something of that living invisible world with which he co-exists — his Earth-born companions and fellow mortals with identical metabolic processes to his own and presumably common ancestry — and challenged this narrow concept of himself as Heaven-sent proprietor of all the Earth.

Now, realising that he is parasitic, destructive, numerically small and with a still fragmentary knowledge of the world he claims to rule, he might fear that he is of less importance in the order of things than the microbes on which the fertility of the Earth depends. To whichever source, divine creation or natural evolution, he may attribute his own origins, he cannot escape certain biological facts the living cells reveal. He cannot show that the twenty amino-acids which build the proteins of his own body differ from those in the lowly microbe, nor that the four nucleotides which form the basis of the hereditary substance of the bacterial cell are not precisely the same as those in his own. The protoplasm, the collodial jelly which is the physical basis of life, fills alike the cells of large and dominant animals like himself, and the cells of viruses that have a volume measured in thousandths of a cubic micron. Both need proteins, carbohydrates, fats, fluids, minerals and oxygen.

Deprived of vitamins, man and microbe both sicken and die. Each journeys from birth through youth, maturity, and age to inevitable death.

Truly, here is an unsuspected brotherhood!

These unpalatable facts he has not readily accepted. Little more than a century ago, men like Semmelweiss in Hungary and Pasteur in France, were ridiculed by their scientific colleagues for claiming that living invisible organisms were the causes of such diverse problems as deaths in maternity hospitals, undesirable fermentations in wines, and the decay of the silk industry because the silkworms died before they could produce their dainty golden threads. When the truth of these observations could no longer be denied, the ancient beliefs in vague environmental influences, slowly gave way, and the bacteriological laboratory began to lead Medicine, Agriculture and Industry into a new appreciation of biological science and of those unseen inhabitants of a world of which Man had known nothing.

Discovering their fatal part in the pestilences which destroyed him, Man characteristically set out to destroy them instead.

Now that he could identify these microscopic miscreants which had attacked him, secure in their invisibility, he would eradicate them all, viruses, rickettsia, bacteria and protozoa, from his Earth!

But they were not eradicated and Man is now learning the many complicated means by which his wily adversaries avoid destruction; their powers of genetic variation and the surprising speed with which the obstinate microbe can change its chemical structure to survive his dangerous antibiotics.

As his knowledge grows too, Man is realising that these organisms are mostly his loyal allies. That he is just a small fragment of Nature's social union. That the invisible members of the Earth's swarming life feed him, clothe him and purify his garbage-laden world and his polluted air.

That they are his constant friends, even though they may at times unwittingly kill him.

THE TWO WORLDS

She trimmed the lamp and made it bright,
And left it swinging to and fro
While Geraldine in dreadful plight
Sank down upon the floor below.

Christabel
Coleridge.

Watching a lamp which the lamp-lighter had left aswinging in the Cathedral of Pisa in 1581, a young Florentine who had been sent by his father to study Medicine at the University of Pisa noted that the time taken for each oscillation was always the same whatever the distance travelled, and from this fact, he enunciated the principle of the isochronism of the pendulum. Realising his deficiency in mathematical knowledge he then persuaded his father to allow him to exchange Medicine for Mathematics and thereafter during a studious lifetime he contributed to Science the hydro-static balance, the principles of dynamics, the mathematics of projectiles, improvements in the telescope, facts about the nature of the Milky Way, about Venus and Saturn and the daily and monthly variations of the moon.

He discovered sunspots and the satellites of Jupiter. He recognised and made use of the Laws of Force and Motion and so provided Sir Isaac Newton with some of the basis of what came to be known as Newton's Laws. He described the hills and valleys of the moon, proved that the Earth rotated on its axis, and demonstrated the error of the accepted doctrine of Ptolemy, that the Earth is the centre of the Universe, and the truth of the heliocentric theory of Copernicus that it is not. Though he did not invent the telescope, he so improved the primitive Dutch invention that he made it into an instrument to explore space, and with it he opened to human scrutiny, the world of the infinitely large — uncountable millions of heavenly bodies spinning in immeasurable space, in distances which are quite inconceivable.

His name was Galileo Galilei.

About 40 years after Galileo died, having taught Man to look outwards from his earthly habitation, to other worlds, another Dutch spectacle maker in his small shop in the City of Delft, having likewise observed that a primitive arrangement of lenses with which he was experimenting had the power to magnify small objects, scraped some tartar from his teeth, and thus examined it. He noted that it contained small forms of life, the movements of which were "very pleasing to behold" and he likened them to "a great number of gnats or flies disporting in the air". He had discovered the world of the infinitely small and of these two worlds, his was perhaps to give the greater benefits to the living things of Earth including Man himself.

His name was Anthony van Leewenhoek.

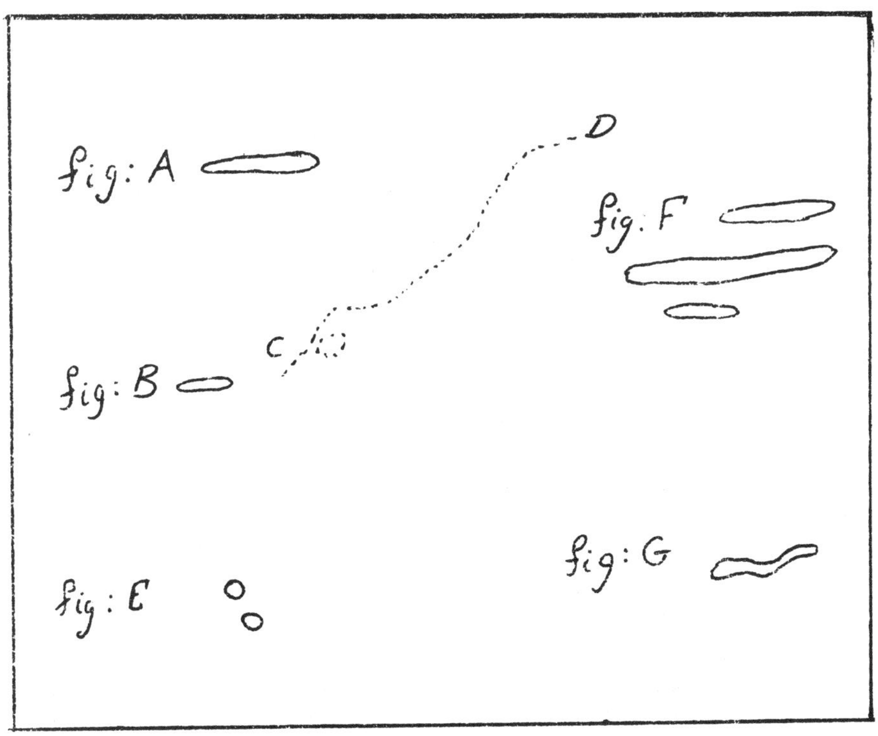

Leewenhoek's record of organisms in pond water.

Astrology and Mathematics had belonged to the ancients whose familiarity with the heavenly bodies grafted to the mathematical knowledge of the earlier civilisations, gave birth to Astronomy. There had been many astronomers and mathematicians before Galileo and men at once appreciated the value of his work. But no one before Leewenhoek had seen the swarming life in "that undiscovered country" which had now revealed to him some of its more massive inhabitants. His papers such as those on the shapes and movements of bacteria and spermatozoa, on yeast cells and the structure of muscle fibres were applauded by the members of the Royal Society as interesting curiosities, and forgotten for 200 years. Learned men could ponder the immensity of space, but they could not conceive the smallness of the cell, the molecule or the atom, or contemplate the invisible.

While mathematical reasoning is clear and precise, biological studies are often disappointingly inconclusive. But the living substance of man or microbe, the lightning swiftness of the chemical changes which give energy to the wings of a gnat, or sight to the eye of an eagle — who can write an equation for these:

"What Immortal hand or eye,
Could frame thy fearful symmetry"

is as true of the virus particle as of the tiger.

And Gallileo would have been amazed to know that the inconceivably small electrons, in the inconceivably small atoms in the inconceivable smallness of the living cell are in motion too in like orderly manner to the heavenly bodies speeding through immeasurable distances in the immeasurable immensity of space.

1

MALARIA

This day relenting, God
Hath placed within my hand,
A wondrous thing, and God
be praised. At His command,

Seeking His secret deeds
With tears and toiling breath,
I find thy cunning seeds
O million-murdering death.

Sir Ronald Ross
Indian Medical Service
Calcutta 1898.

Hymn written in thanks on discovering the method by which mosquitoes transmit malaria, after years of patient research.

The causes of human malaria are protozoan parasites known as Plasmodia, a species which employs an unusually rapid method of multiplication by which the nucleus of the cell divides into many separate nuclei and the organism then splits up into as many new organisms as there are new nuclei. In this way they multiply asexually in Man but they have also a complicated sexual life cycle in some species of mosquito.

There is no medical term more romantic than malaria and no other malady whose incidence and investigation so nearly bridges the whole span of man's recorded history. The very name 'mal' 'aria' carries us back from today's sophisticated laboratory studies of the physiology of the parasite, to ancient theories of malevolent environmental influences

— the evils of the night air. Between the two there are centuries of patient study of the parasite and its vectors, the world wide mosquito.

Hippocrates in 500 B.C., making the first attempt to systematise the scanty medical knowledge of his day, described the clinical symptoms and gave names which are still used today, to the various forms of malaria, yet it is only in recent years that the mystery of the temporary disappearance from the blood of the newly injected parasites has been cleared up by proving that they first migrate to the cells of the liver to multiply their numbers.

With all our knowledge, however, the plasmodium of malaria defeats us still. The hygienist has sprayed the mosquitoes marshy breeding places, the engineer has drained them, the doctor has given preventive drugs to half the Earth's exposed peoples, yet the yearly cases of malaria are counted in millions. The weight and scientific power of the World Health Organisation, which has at last freed the World of smallpox has failed to do the same with this dainty Protozoan which man and mosquito together have sponsored through immeasurable time. The sudden fever, the distressing chills, the violent headache, the abdomen distended by swelling spleen and liver, the intolerable weariness of malarial anaemia, the rapid deaths of the malignant form, or slower of the more benign, which destroyed former civilisations to leave only ruined homes, and fields abandoned to the wilderness, were just the same as those which today overtake the inhabitants of idyllic tropical islands or hot and steamy jungles.

The incredible life cycle of these organisms, so minute that a family of them can dwell together within a red blood cell, itself so small that it is counted in millions in every millilitre, involves intricate and dangerous journeys from the human blood where they are multiplying asexually, to the mosquito where they unite sexually and then back again to their former hosts to multiply asexually again.

Pumped into the blood stream with the saliva of a feeding female mosquito, they first seek safety in the liver cells from the blood phagocytes which would destroy them. Each parasite multiplies there by a simple process of division to thousands which rupture the cell by their very volume, and then each of these enters either another liver cell or a red blood cell and continues this process of multiplication to many millions. Like destructive vandals they smash down their adopted home every few days and squat in another. But no short-cut methods can replace the sexual process except for brief intermediate periods, and from amongst these millions of sexless parasites travelling round the body in the red blood cells and dividing up to start their families in a ''parents-without-partners'' relationship, a few change into males and females by some incomprehensible adaptation.

They alter slightly in form and structure, they no longer multiply and shatter the red cell but quietly await their destined end — a honeymoon in a mosquito's stomach.

Do we see here the origins of sexual life? Is this how all males and females arose?

These astonishing parasites whose life is split into two, a sexual life in the mosquito and an asexual life in human beings must hold the answers to many mysteries. How did they first effect this improbable partnership of man, mosquito and malaria parasites? How do they change non-sexual to sexual cells?

Having made the change, why will they not unite in the blood of man, but only in the mosquito's gut?

These are things we should like to know. But, come to think of it, why should we probe the private lives of everything on earth? We have problems enough of our own to which we need to find the answers — narcotics, alcohol, automobiles, war, greed, riches and poverty — which plague the human race in a way the little plasmodium of malaria never did.

"OF HIS BONES ARE CORAL MADE"
THE TEMPEST

I must go down to the sea again, to the lonely sea and the sky,
And all I ask is a tall ship and a star to steer her by;
And the wheel's kick and the wind's song and the white sails shaking
And a grey mist on the seas face, and a grey dawn breaking.
I must go down to the sea again, to the vagrant gypsy life,
To the gulls way and the whale's way, where the wind's like a whetted knife,
And all I ask is a merry yarn from a laughing fellow-rover,
And a quiet sleep and a sweet dream, when the long trick's over.

Sea Fever
John Masefield

What happened to La Perouse?

On August 22nd, 1741, in a small village near Albi in France, Jean Francois De Galaup La Perouse drew his first squalling breath.

When did he draw his last, and where?

He had been sent from France to investigate the fur trade in North America and the prospect of whaling in the Southern Ocean and to look for the North West passage.

In January 1788, he called in to visit the newly established English Colony at Botany Bay. These exiles, forlornly planted in a desolate continent which they feared and hated and yet did not wish to share with any other nation, met him with distrust, for National rivalries persisted between these ancient adversaries. On February 7th, La Perouse sent a letter to his Government and sailed away to eternity.

Nearly forty years later the wrecks of his ships were found submerged in the clear water of a beautiful coral reef near the New Hebrides.

What mischance put them there and what happened to the captain and all his men?

Facing the harbour at Vila, in the New Hebrides, there is a museum which contains some pathetic relics from the wrecks. Some coins, a sailor's shoe of the 18th century with a big silver buckle, some carpenter's tools, and, on a concrete base on the lawn, the large rusty anchor which once slid down from one of La Perouse's ships to give it rest and safety on the surface of many waters. Hand forged in the naval yards of pre-revolution France from iron, mined, smelted and wrought by the labour of forgotten men, the long anchor adorned the prow of BOUSSOLE or ASTROLABE, as the two ships started on the journey designed to bring wealth and perhaps new lands to France.

It had dropped to rocky anchorages off Alaska, had gripped the muddy harbour bottom of Vladivostok and the warm sands of Botany Bay, which honours its captain with a monument and has put his name to a great industrial suburb. It had danced across the blue Pacific, over the grey seas of Asia and the intense ultra-marine of that lovely tropic bay, where it had left the name of the ship it probably served for future centuries to remember — Astrolabe Bay.

It came at last, to lie amongst those fantastic corals which can rip the modern steel hull as easily as the wooden planks of 18th century ships.

Now it rests on New Hebridean soil, large rusty flakes of iron slowly returning it, as its former masters have returned, to the Earth whence it came.

Perhaps the ancestors of the people who live not far from Vila, knew something of the fate of "Boussole" and "Astrolabe" but no legends circulate here which might have been inspired by shipwrecked Frenchmen, though Captain Cook and Bougainville are remembered.

These may be the tropic isles of paradise, of blue seas, sunshine, sandy beaches, pandanus and palm. They are also the isles of mosquitoes and malaria, hookworm, filaria and ascaris, taboos and black magic. The incantations and crafty arts of the medicine men, though of undoubted psychological value to the afflicted, do not exorcise the plasmodium of malaria from the blood or drive the vampire hookworm from the bowel. These islanders know that while they sleep, evil spirits steal a little of their blood and so bring curses and illness on them, for generations of medicine men have told them so.

And the medicine men are right.

At night, on the warm Pacific beach or in their low dark community huts when the smoke from their smudge fires has cleared, these demons, all female, big with fertilised eggs and driven by the gruesome urge to drink human blood, fly down in the late twilight to take their vampire meal.

Within the next eight days she needs two more meals of human blood to mature her eggs and then, alighting on the still marsh water she expels them one by one, till about one hundred, boat-shaped and provided with floats, are left swinging together to form stars and triangles on the water's surface as though to hide their evil purpose with celestial designs.

Following the patterns of an evolutionary eternity, each egg hatches first into a water-baby which lies suspended under the water surface, breathing through its tail. Then, as though acknowledging a mistake, it changes shape and now breathes through snorkels in its head. Still unsatisfied, it takes the parents form and breathes through perforations in its thorax, then breaks out of its unsuitable skin which it now uses as a raft while it spreads its newly grown wings to dry.

Though she has slaughtered men and women in millions, killed the foetus in the impoverished womb, soured the happiness of infected childhood, defeated armies, unseated Emperors and made deserts of fertile fields, she has been a close partner of the human race through immeasurable time. She and man together, the cold arthropod and the warm blooded vertebrate, have undertaken the nurture and propagation of the plasmodia which cause human malaria, for she is the Anopheline mosquito.

Man is both her accomplice and her victim. In his malarial blood there circulate the sexual malaria cells which only the mosquito can activate. She will incubate them into the asexual cells from which only Man can grow sexual cells again.

Standing on her head, her straight body at an acute angle to her victim's velvety skin, she draws his warm parasite-filled blood into her cold gut, and the magic of sexual union begins.

Most living things are either male or female and their union produces new individuals.

While most of the malarial cells in human blood multiply continuously in a manner which does not involve sexual differences, others deviate from this non-sexual course to become gametocytes or sex cells.

Though now male and female, they circulate side by side for about ten days uninfluenced by the universal attraction of the sexes and then prefer to disintegrate and die rather than to unite and produce a new generation, in this most noble medium — this human blood.

(continued on page 8)

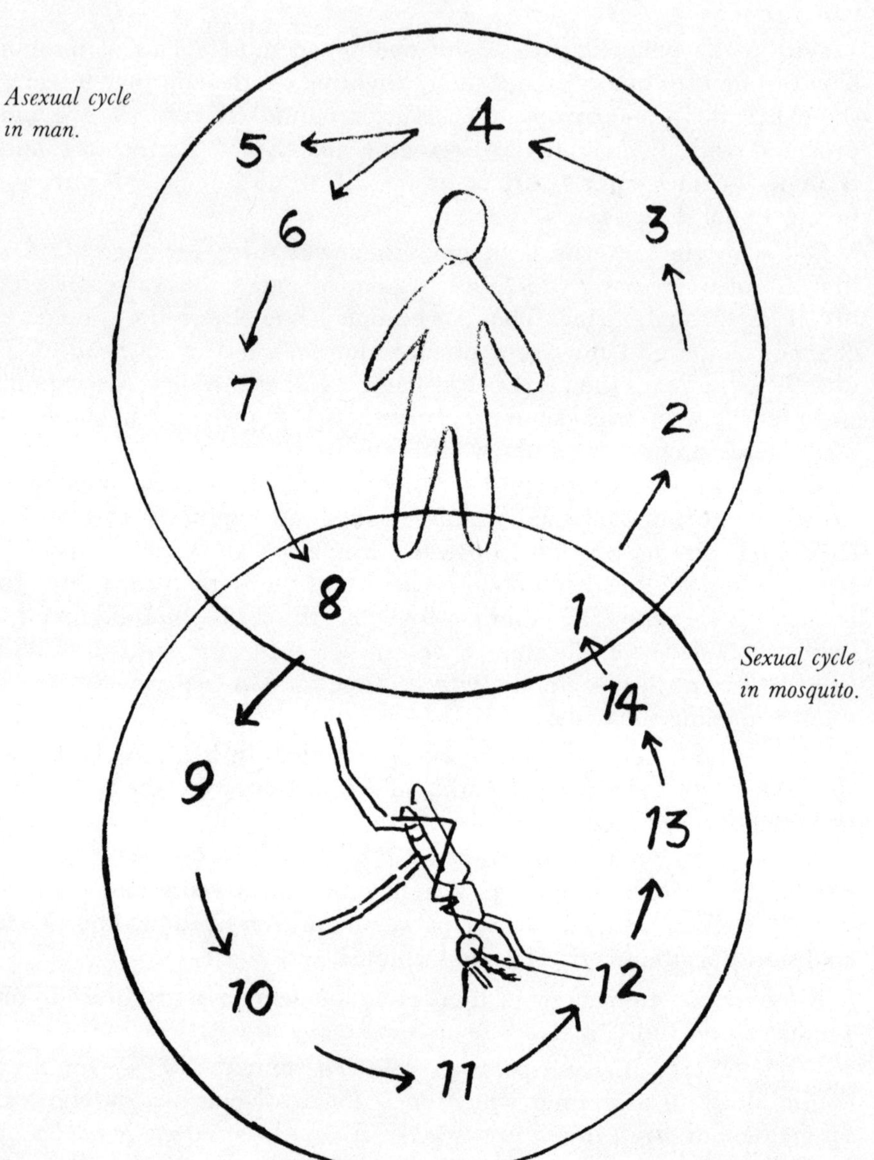
The Malaria parasite.
Asexual cycle in man.
Sexual cycle in mosquito.
1
2
3
4
5
6
7
8
9
10
11
12
13
14

THE LIFE CYCLE OF THE MALARIA PARASITE

1. *About-3000 asexual sporozoites 15 microns long injected from salivary glands of biting mosquito.*

2. *Sporozoites circulate in human blood.*

3. *Sporozoites enter liver cells and multiply to some thousands — 10,000 to 40,000 have been demonstrated.*

4. *Distended liver cells rupture and parasites now in rounded forms called merozoites escape.*

5. *Some are destroyed by the white defensive cells; while some enter other liver cells and multiply again.*

6. *Some enter red blood cells and multiply to 8, 16 or 32. Assuming "ring" forms and now known as trophozoites. The distended red blood cell ruptures and releases these trophozoites into the blood stream. They are still all asexual.*

7. *These released trophozoites enter other red blood cells repeating the sequence of multiplication, distention and rupture.*

8. *Some trophozoites enter red cells but do not divide, instead they become sexual cells — male and female gametocytes and circulate in the blood for ten days when they disintegrate if not imbibed by a mosquito.*

9. *These sexual cells are drawn into the mosquitoes gut with human blood.*

10. *They unite sexually.*

11. *The now fertilized female cell implants itself in the gut wall of the mosquito.*

12. *It segments into numerous asexual sporozoites, distends and ruptures as the other cells have done.*

13. *The sporozoites invade the whole body of the mosquito including its salivary glands.*

14. *The mosquito bites a human being and expels its saliva to assist penetration of the skin, thus carrying sporozoites into the blood stream.*

Patrician or plebeian, King or peasant, master or slave, white or coloured, victor returning in triumph from the wars or vanquished led in his captive train, all may be full of the parasites of malaria, but the warm blood of none of them will sustain the sexual activity of these minute offshoots of the Plasmodial life cycle — these gametocytes, these marriage cells.

But let them be sucked into the cold stomach of the fecund mosquito, and within minutes they are flung into an ecstacy of reproductive change. They batter down the walls of the red cells which have harboured, nourished and transported them, and lie free in the stomach cavity. The female cells make some changes in their nucleii and expel some of it in expectation of fertilisation, thus coming to resemble the egg of the female of higher species from which men and mice arise.

The male gametocyte is not so placid. In a fury of activity, his nucleus divides into about eight fragments. From each a lashing thread-like flagellum, twenty-five microns long, quickly grows and they are then thrown out of the parent cell. Instant spermatozoa! Now the race is to the swift. One of them first reaches the waiting female cell, sheds the flagellum which has powered it to its goal, enters the cell, and their nucleii fuse into one. This fertilised female gametocyte — this fertilised ovum — this fertilised egg — for the processes are the same as in those animals we call "higher animals" and the relative sizes of male to female gametocyte is as the sperm to the ovum, now implants itself in the stomach wall of the mosquito just as the fertilised ovum implants itself in the wall of the uterus, to commence segmentation into new living beings. In about ten days time, distended to bursting with several thousand new forms each 15 microns long, known now as sporozoites, which are the children of this fertilisation, it ruptures and

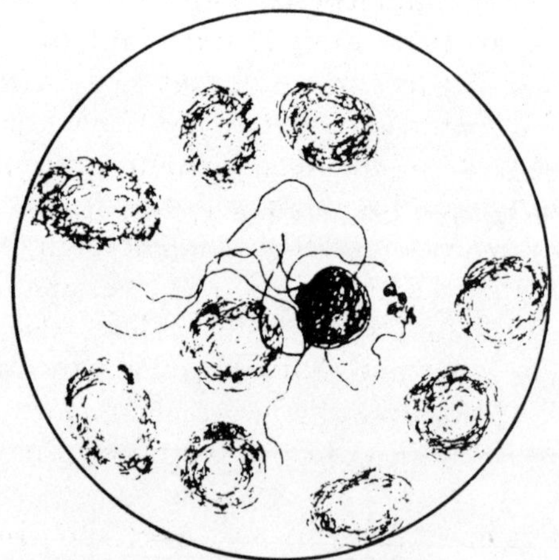

Male malaria cell amongst red blood cells in mosquito gut.

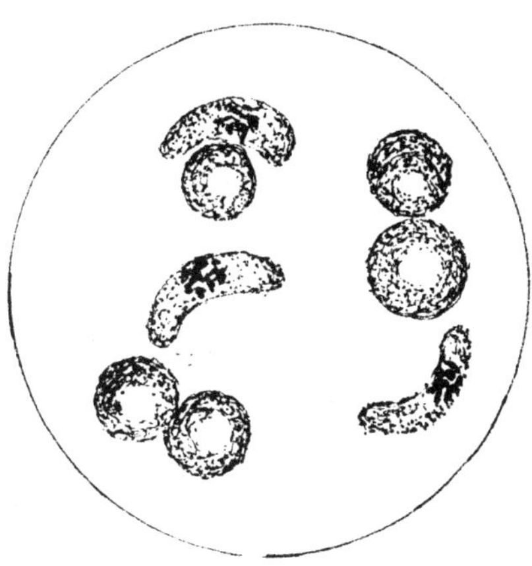

*Gametocytes of malignant tertian malaria
amongst red blood cells in blood stream.*

releases them to infect the mosquito in every part including her salivary glands. She is now a malaria carrying mosquito ready to play her part in the eternal cycle. But it is to be noted that *MAN* has infected *HER*. Until she imbibed, with his parasitised blood, these gametocytes which only he could develop, and grew them into the sporozoites which none but she might generate, she was a harmless insect.

Only the parasite benefits from this unnatural partnership. It is man who suffers and dies. It is the small child of these tropic isles who sits languidly anaemic, with distended spleen. It is the one million people whose lives each year are forfeited to the plasmodium and the 300,000,000 people of the Earth who yearly suffer its acute attacks. It is the expectant mothers of the tropical world whose parasitised blood aborts their quickening young. It is the soldier, the prospector, the villager, who lie sweating and fevered in malarial paroxysms. It was the opposing armies of Britain, France and Germany whose hostilities ceased in common suffering when this common enemy invaded their warlike blood in Macedonia in 1916. Was it not the same defenders who met the mighty hosts of Persia on the same ground and made them abandon their campaigns just short of victory when they came to subdue the haughty Greeks? Did these same foreign cells, slave-carried to Rome's highly civilised blood, allow the Goths and Vandals from the cooler north to win a barbarian victory? Did Alexander the Great sweep across Asia to India and destroy the Persian Empire because he had left his main enemy behind in Macedonian swamps? Did the power and civilisation of ancient Ceylon crumble under the attack of a mosquito?

Two thousand five hundred years ago, when the Persian Empire was founded with Persepolis as its capital, Hippocrates described the fevers of malaria giving them the names we use today — tertian, quartan and quotidian, and noting the enlarged spleen and liver. Persepolis now lies in magnificent ruins but the plasmodium has changed not at all. Through millions of transfers and incomprehensible millions of parasites, it has maintained its purity of line.

Recorded history though is but a speck in Time and we have no knowledge of what went before. Sometime in unknown geological ages, these plasmodia first must have entered human blood. Somehow they developed the biological urge to split those male and female gametocytes off from their routine asexual cycle, for the mosquito to suck out and hatch into sporozoites. Somewhere these cunning seeds of malaria had learned to co-operate with Man and insect in complicated cycles for their survival, though Man died in his millions to serve the purposes of a microscopic blob of protoplasm.

Where then is the beginning of this revolving story? The mosquito is infected by the malaria sex cells in her human blood meal from a person previously infected with malaria non-sexual cells by another mosquito. She hatches them, in her turn, into non-sexual cells and pumps about 3,000 of them into a new victim. They disappear within the hour, to hide within the cells of the liver, and each one which enters a liver cell, divides during the following week, into thousands, this time without the aid of male or female for only in the mosquito does sexual union occur. These over-distended cells rupture and release the parasites. Attacked by white blood cells, they seem to realise the need for further reinforcements, though the 3,000 has already become many millions, so some re-enter other liver cells and multiply by thousands again. The others enter the red cells and also multiply until the total number reaches about 40,000 in each microlitre of blood. Then, a select few of this number become male and female cells to start the process of regeneration again, for which they too must take their dangerous way from man to mosquito and back to man again. What hormones do they invoke? On what royal jelly do they feed that these cells, stemming from the same source and circulating in the same blood stream as their fellows, take on the form of male and female and assume responsibility for procreation and continuance from the eternity of the past to the eternity of the future of their micron-measured brood, while the others only find their normal way to multiplication and destruction?

Now that man has uncovered the villain will he destroy it?

It is estimated that if he could, he might suffer still greater calamities; that the removal of this biological control to population would be followed by such an increase that famine and wars for survival would destroy the society Man has developed.

Such ambitions must await other sociological progress, but like the mystery of the plasmodial life cycles and the mystery of the fate of La Perouse, it is something to which Man may never find an answer.

MALARIAL MATHEMATICS

The blood volume of a 75 kg man is approximately 6 litres.
Each litre contains 1000 millilitres.
Each millilitre has 1000 cubic millimetres.
Each cubic millimetre contains 5 million red blood cells.
In the blood of a 75 kg man therefore there are—

$$6 \times 10^3 \times 10^3 \times 5 \times 10^6$$
$$\text{or } 6 \times 5 \times 10^{12}$$
$$\text{or } 30,000,000,000,000 \text{ red blood cells.}$$

The upper limit of infection of red cells by the parasites of malignant tertian (MT) malaria is 10% or 3,000,000,000,000.

The upper limit for benign tertian (BT) malaria is only 1% or 300,000 million red cells.

2

POST PARTUM FEVER

Any feverish condition occurring in a woman within twenty-one days of having a baby or a miscarriage is named puerperal fever and it can include such conditions as breast abscess or bladder infection as well as those dangerous invasions of the blood which proved so often fatal — eighty per cent fatal — before the use of penicillin.

Besides the normally harmless organisms living in the throat, the nose and the intestines of the patient or her attendants, certain dangerous little fellows, spherical in shape and one micron in diameter, who look either like bunches of grapes or long bead necklaces when they grow and form colonies, and who are known as staphylocci and streptococci are the ones which produce severe infection and death within a few days.

Everyone working in or near maternity or labour wards is now screened for the presence of these and if positive kept away until swabs taken from them are negative. Now childbirth has become just a normal joyful event in a woman's life, almost completely free from danger or fear.

To the midwife of the past, today's sterile maternity hospitals would have been a hopeless puzzle. In place of a clean sheet or two in the patient's home, the wash-basin on a tripod in the corner, and of course her obstetric bag, brought direct, most likely, from the dusty floor of the horse-drawn vehicle she has travelled in, there is cleanliness in the tiled walls and floor, the chromium fittings, the snowy gowns and caps and masks, the piles of linen, the disposable gloves, the canvas over-shoes, the carefully covered instruments, the antiseptic dispensers, the automatic taps and wash-basin, and safety in the modern anaesthetic machine and in the transfusion apparatus and the humidicrib ready in case of need.

But as these furnishings have come into maternity wards, death has departed. From a mortality rate of about 50%, which was the loss in one hospital in Vienna in the 17th century, it has been reduced to

"

vanishing point by clean and gentle techniques, correct pre-natal care and latterly the use of antibiotics to combat the occasional infection which occurs in spite of asepsis.

Even the maternal home was safer than the wards of the lying-in hospitals where infection was so often carried by the staff from the sick and dying in other wards to the young mothers in theirs, or by doctors and their students, even from the post-mortem rooms.

Though we can now be proud of our obstetrical records, these results could have been achieved nearly a century and a half earlier if those thinkers who suggested methods of preventing deaths had been listened to. Charles White in Manchester, in 1773, Alexander Gordon in 1795, Ignatius Semmelweis in 1847 and Louis Pasteur in 1879 suggested that microorganisms may cause the deaths but only in 1935 when Rebecca Lancefield pioneered a method of typing microorganisms which enabled her to trace their sources with precision, could they be positively incriminated.

IRISH EYES AND IRISH APPLES

Season of mists and mellow fruitfulness,
Close bosom-friend of the maturing sun,
Conspiring with him how to load and bless
With fruits, the vines that round the thatch-eaves run,
To blend with apples the mossed cottage trees
And fill all fruits with ripeness to the core.

To Autumn
John Keats

'Tis but a step down yonder lane,
And the little church stands near,
The church where we were wed, Mary,
I can see the spire from here.
But the graveyard lies between, Mary,
And my step might break your rest
For I laid you darling down to sleep
With your baby on your breast.

The Lament of the Irish Emigrant
Lady Dufferin

Liverpool is a big and busy port. A grimy, foggy, cold and rainy port. A cheerful, happy, and modern port. A port with nearly thirty miles of quays and about five hundred acres of docks and basins.

Visited by the ships of all nations in peace, in war it is the British end of the Atlantic life-line and the hub of the railway systems which disperse to all parts of the British Isles the goods it receives from all parts of the world. It has handled an enormous tonnage of American cotton consigned from the plantations to the mills of Lancashire. Its great fleets of refrigerated vessels bring the bulk of Britain's meats from Australia and the Argentine, and the amount of grain which pours into Liverpool makes it the world's biggest grain market outside the United States of America.

As John Carson's ship edged its way up the Mersey to berth, it was not of Liverpool that John was thinking, but of Portadown across the Irish Sea where Sarah Lindsay was waiting to be his wife. He was leaving his job as a marine engineer to buy his father's hedge-enclosed apple orchard, with its ancient gnarled trees and its slate-roofed old white stone house on the Dungannon Road which had been in the family for generations.

He was paid and farewelled at the Company's offices in Lime Street, took the 'bus to Pierhead and the night packet for Belfast. And what a slow journey it was from the wharf through the tedious docks whose designer surely never had a girl like Sarah waiting for him on the other side of the Irish Sea. He pitied the port officials walking along concrete walls in cold driving rain giving megaphone orders to other poor shivering men in oil skins and sou-westers who edged that ship through and into the open sea at last.

They were late at Belfast next morning because of a snow storm. John impatiently watched the trains running along the green shores of Belfast Loch and leaving white trails of water vapour in the cold air, and the skeleton ships in the shipyards where the Loch narrowed to become a river. The city was newly dressed in snow and the road to Portadown dazzingly white. Portadown, now to be his home and Sarah's and perhaps their children's.

A few days later John and Sarah stood together at the altar in the old Drumcree Church where many of their ancestors had been married. It was sad though to read the memorial inscriptions on the tombstones for the brides who had stood one year where Sarah stood now, radiant with love, and who slept next year in the churchyard with an infant son or daughter. "Child-bed fever" they called it then — the fearful risk of matrimony; so often the sad end to a loving union.

As the seasons moved around the world, they brought a fine apple crop to John and Sarah in the following year. The hidden life forces in the apple trees had known when it was Spring. Ask not how they knew. Neither ask how the trees took the minerals and water from the soil and the carbon and oxygen from the air, and with the mysterious help of sunshine fashioned them into leaf and bud and blossom, sap and juice and fruit. Let them grow — big sweet red apples, and leave the mystery of it to the apple trees.

Autumn brought its lovely colours to the orchard again and as Sarah was helping John with the last of the crop, she told him that next year he must manage the work himself for she would have a baby to care for. He kissed her till her cheeks were as red as the apples in the bin and the sunset in the sky. Then they sat beside the unpacked fruit, listening to the future; hearing a baby crying in the nursery, the patter of tiny feet in the house and childish prattle at the fireside. They thought no more of the mystery of the coming child than they did of the mystery of the growing fruit. You took the apples as your due, for had you not cared well for the productive tree? You received the child with love and affection, for was it not a part of each of you in love begotten? Yes, indeed; but behind the well-earned harvest stand the intricate secrets of the living factories of Nature. And in the now developing human-being-in-miniature which Sarah has announced, lie the beginnings of the cells of all the organs which will be his or hers. And some of these will become the centre of the continuing germ plasm of human life able to pass on those cells with the power to create a new human being, with power to pass them on again — something surely resembling the concept of immortality!

"Of Earth's first Clay they did the Last Man knead,
And then of the Last Harvest sowed the Seed."

The days of approaching parenthood went happily by till the time came when, as the apple matures and is cast off by the tree to sow its own seed and grow to another tree, so the new being which Sarah had nurtured with food, oxygen, warmth, with fluids, and all the intimate unknowns of the gestating months was ready too, for independence. With a reversal of those chemical controls, the hormones, which had defended it against expulsion, it was now told to go, and with slow rhythmical steps, heralded by uterine relaxation ahead and pursued by waves of muscular contraction behind, it progressed from the world of shelter and peace to a new world of stress and war, whose chill blast, striking its unfamiliar skin, was met by a protesting howl which opened its lungs to receive the oxygen its mother's blood would no longer supply.

A child was born.

Now, that from which it grew, the placenta and cord, root and stem of its intra-uterine life, must be born too, and the contracting waves peeled the adherent cotyledons gently from the bleeding surfaces. The pituitary gland gave out the oxytocin which contracted the soft flabby exhausted mass of the uterus to the size and hardness of a cricket ball, enfolding the vulnerable site of the expelled placenta within its mushy walls, while its criss-cross muscle fibres clamped down on the mass of severed blood vessels sealing them against further haemorrhage; and Sarah slept in the happiness of motherhood achieved.

On all exposed parts of the body there are micro-organisms, some of which could be harmful were they not controlled by mechanisms which the body has devised for its defence. But in sheltered positions where no micro-organisms usually penetrate, there is no such defence. The area on the inner surface of the uterus from which a placenta has just separated, is not only a normally sheltered area, it is now also an open bleeding wound filled with blood clot and serum which is highly nutritious to many kinds of invading germs. Should these by some accidental contamination reach this wound, they can produce a fatal septicaemia, when, in the same patient's throat from which they may have come, they cause nothing more than mild inconvenience.

Carried there perhaps in the blood from some other part or introduced from outside by some manipulative process, streptococci found their way to Sarah's womb in the first few days of her motherhood. The streptococci are spherical assassins, one micron in diameter, quite immobile, but capable of manufacturing about twenty toxic substances which are able to dissolve the red cells of the blood, destroy the white cells, liquify the cementing material of body tissues and the blood clots with which the body tries to debar their progress. In the broken fragments of these clots, they are washed into the circulating blood and scattered throughout the body. Having thus, by chemical means, made up for their lack of mobility, they were carried in uncountable myriads along with their destructive poisons to infect all the tissues of Sarah's body, which reacted to their unwelcome presence with the signs and symptoms of fatal illness.

When John saw Sarah pale, bloodless, and still, against the pillows he knew the awful fear which had entered the door of many a matrimonial home before his own. His parents reassured him only that he might enjoy what hope he yet could before the blow which they knew so well, fell on him, and his happiness turned to anguish.

His father had once had a favourite sister, Maggie, who like many others they had known had died of child-bed fever. As he gazed sadly into the evening fire, the old man recalled Maggie with her glittering eyes, burning like the glowing peat before him from her high toxic fever. He saw the two bright little red patches in the sallow sunken cheeks. He could feel again her racing pulse, her fevered hand, so burning hot though she was shivering while he spread blankets to cover her chilly limbs. He remembered the agony of her face as the flitting pains stabbed her, and her mind alternately sharpened to an abnormal clarity or dulled by the dreadful malady. He had sat by her bed and watched her go down to delirium and death with the doctors quite unable to save her.

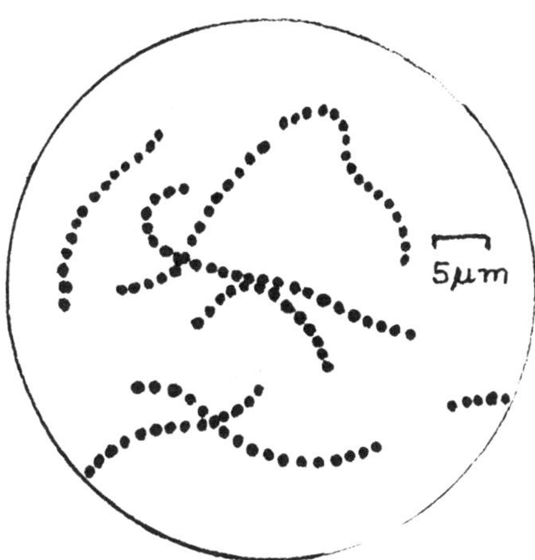

Streptococcus pyogenes. 5 microns on same scale of magnification.

Now, God help us, Sarah has the same trouble.

In a few days Sarah's eyes were aglow with the toxic glitter and high fever of septicaemia and her sallow cheeks concentrated their colour in two bright red unnatural patches. She was burning and chilled, in pain, alternately unresponsively dull or unusually alert. The chains of swarming streptococci, the tiny spherical microscopic globes just one invisible micron in diameter, were killing her as surely as they had killed Maggie and other new mothers, and there was nothing but gloom in the orchard and beside the hearth fire.

John was late coming in on the eighth evening and his parents sitting despondently on either side of the ingle, thought that this meant Sarah was worse. When they heard his approaching footsteps at last, they looked anxiously towards the door where he entered quietly.

"Sarah is better tonight", he said, crossing the room to warm his hands at the fire.

"Better?" asked the old man incredulously. "Did you say better son?"

"Aye, Father, better. She has had no fever today. The doctors have given her something new. They call it penicillin and they say it is saving all these mothers now."

"God bless it whatever they call it", said his mother, weeping on his shoulder as the dark cloud that had hung over her and all parturient women, seemed at last to show some promise of dispersion. She had been no stranger to the terror which so often walked into the lying-in hospitals with the happy young wife, hoping there to bring forth a new life and finding death instead. Like all other women she had entered each pregnancy with fear and emerged with thankfulness.

"God bless it, son if it has saved our Sarah."

3

TUBERCULOSIS

Tuberculosis is the chronic progressive condition produced in almost any organ, but chiefly in the respiratory tract of man and animals by the moderately toxic bacillus, Mycobacterium tuberculosis. They are fine rods about 4 microns by 0.4 microns, having a waxy content which makes staining difficult, but when suitably stained they retain aniline dyes in the presence of acid-alcohol and are described as acid-fast, a characteristic also present in the bacilli of leprosy. They are difficult to grow in the laboratory and need a full oxygen supply and specially enriched culture media.

When the wealthy Egyptians entrusted their bodies to the skilful arts of the ancient embalmers for mummification after death, they also entrusted some of their most personal secrets with those bodies, to be revealed centuries later by the skill of the anatomist.

Syphilis, for example, leaves its mark in bone deformities and corroded arteries. The deformities of the leper or the arthritic are as evident in the dusty corpse of antiquity as in the living patient of today. They can speak to us of a thousand years ago, and show that the flesh of the ancient aristocrat in his luxurious palace was heir to the same ills as those which affect mankind today in palace or in slum.

The tubercle bacillus leaves the scars of its ravages not only in the lungs but also in the collapsed vertebrae of the "hunchback" spine. In the enlarged tuberculous glands and fibrous tubercles we may see the characteristic groups of defensive cells which tried vainly to save the dying Egyptian.

As with the bacilli of leprosy, the waxy outer coat of the tubercle bacilli resists the phagocytic cells of the body, and both diseases are therefore difficult to cure. The word "phagocytic" literally means "eaters of cells" and these white defensive cells which circulate in the blood and wander like blood-hounds through the tissues actually

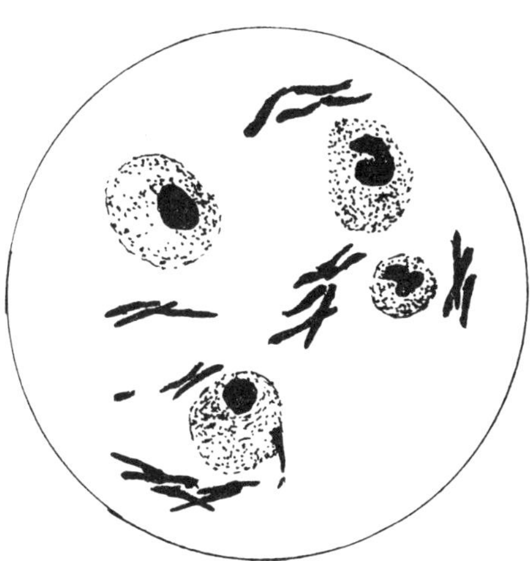

Mycobacterium tuberculosis.

envelop, and digest foreign substances such as invading bacteria. They surround and attack the bundles of turbercle bacilli in military fashion.

Other defensive cells come into the battle — the lymphocytes, the moncytes, the fibroblasts, and the giant cells which are the ultimate in aggressive body resistance, and the stale-mate battle goes on for years revealing itself in the hard swollen glands.

The tubercle bacilli have no known dangerous toxins such as the diphtheriae possess; they are simply squatters who defeat all attempts at eviction, and here as in the case of the trypanosomes of African sleeping sickness, we may see an over-action of the body to a fairly harmless tenant.

The result of this activity is the gradual growth of small nodules, which, if examined microscopically, are found to present an excellent picture of the long drawn out conflict, the same cellular patterns in glands removed from today's patients as from the crumbling Mummy. The tubercle bacilli still occupy their original central positions, and all around the periphery to make escape impossible, are the highly active monocytes, the antibody-producing lymphocytes and the tough elastic fibroblasts. The whole distressing lesion is now known as a tubercle and the disease as tuberculosis for these will now be scattered about the body. The resisting cells have not killed the enemy but they have placed it under seige.

Gradually the tubercle bacilli and their attackers, cut off together from out-side supplies of blood and oxygen die and decay and a cavity is formed within the tubercle. As these central cavities extend and numbers of them join together to form larger cavities the patient becomes gravely ill. But often the body, if it is healthy enough and well fed, can actually destroy this unwanted little bacillus and instead of a cavity, a scar of thick fibrous tissues may form. Although it is still one of the worst infectious diseases with an estimated fifty million sufferers today, the better feeding and housing of the world's population is the most effective weapon against it. Only in those countries afflicted by poverty, hunger and war is tuberculosis increasing.

WAR AND PEACE

"The little mermaid stretched out her arms to the sun, and for the first time in her life, tears moistened her eyes"

The Little Mermaid
Hans Christian Anderson

Christina stood in the Town Hall Square — the Raadhusplads — in knee-boots, macintosh and waterproof hat waiting for her father and the cold, cold rain of Copenhagen's winter beat down on her. It beat on the bronze statues of the Lur-blowers, those sturdy Danes, on their high pink granite pillar with the Lur curling upwards from their mouths so naturally, she almost felt she could hear their music but for the traffic and the rain. The Lur is used now only as a trade mark, an ancient Scandinavian symbol for Danish merchandise.

Father was late; he had been called in conference to hear some of the dangerous news coming from other parts of Europe. Christina walked across the Square amongst the ice-cream and flower vendors who looked so forlorn and unwanted in the drenching rain, and took shelter in front of the Royal Palace. Father had brought her from Odense on her first visit to Copenhagen and he was showing her the sights of the beautiful city from their cosy quarters on the Platanvej. They went to the fish wharf where the trawlers came in from the North Sea with cod and herring packed in ice and saw the marble troughs where the live fish swam to be scooped out, all fresh and flapping, by the fishmonger. Here too, were the fish-wives dressed in long dark-blue dresses with white aprons and white starchy bonnets. Copenhagen had honoured the fish-wife by erecting a statue to her. There she stood on her own

---

FOOTNOTE:
The Lurs were wind instruments from the Early and Late Bronze ages — 1100 B.C. to 400 B.C.

The form of the Lur was a long thin S-shaped horn with the S twisted into two planes more or less at right angles and a mouthpiece which resembled the head of a striking cobra. Some of these recovered from peat bogs in Norway, Sweden, Denmark and Northern Germany, the only places where they have been found, are in such preservation that they can be played. They give out a booming sound which can be heard over long distances. They have been found placed together symmetrically in pairs in the bogs, indicating a probable religious or memorial reason.

The Little Mermaid.

quay, in time-defying sandstone, complete with bonnet, apron, and basket of fish, the stolid, honest symbol of all fish-wives, the world over.

They admired the green curled copper spires of the churches. They climbed the Round Tower, up and up the spiral roadway to its top where Czar Alexander had driven his horses and chariot. They went to the Tivoli Gardens which people came from all the world to see. They sat in the park near the statue of Hans Christian Anderson, and that legendary statue — the Gefion Fountain with the streams of water blowing from the nostrils of the great bronze oxen, as they strained at the heavy plough — while father told her the story of the ambitious Goddess-Mother who had turned her six sons into oxen to plough the island of Zeeland from the land of Sweden and leave a lake in its place.

They went down to the harbour to watch the ships of all the seas, and little graceful Langelina on her granite boulder, with her pretty bronze shoulders and girlish bosom shining wet with rain or spray or glowing warm in the sun, peering down into the water, a dear little timid real mermaid, so lost in thought that she hadn't noticed the beautiful city that had grown up around her.

It was too wet today to do any of these things, so when father came they went to the great railway station where the beautiful candelabra hanging from the high arched roof and making up with their electric brilliance for the dull gray darkness of the winter day, were magnificence indeed to the country child. There was carnival in the hurrying crowds, the trains arriving and departing, some even from Odense, their home and Hans Anderson's.

In a cosy corner of the station dining room as they sipped hot chocolate, father told her how grave was the position of Denmark and its small neighbours. Hitler's armies were on their borders and he had been told that the order to advance was about to be given and the old Danish life soon would disappear beneath a tide of military terror, but he hoped he and 'Tina would always remember with happiness the holiday they were having together now.

When Hitler's tanks came out of the fog across the border, many good Danes besides Christina and her father grieved for the Denmark that was going under the irresistible military power to begin the 'occupation' years of misery, slavery, tyranny and want.

In those crowded damp unventilated houses on a food ration of less than maintenance; in clothing which though worn could not be renewed; working long hours and persecuted even for their thoughts, the Danes knew the bestiality of modern total war. The gold in their banks, the grain in their storehouses, the cattle in their fields, the art in their galleries, the labour of their hands, anything of value, was requisitioned to feed the hungry war machine or to enrich the salons of the victors. Danish lives were expendable; the dead needed no food or clothing. Survival seemed impossible, but Christina survived though she never saw her father or family again.

On May 5th, 1945, the liberating armies marched into Denmark where conditions, as information coming out of all occupied countries by way of the underground network confirmed, had increased disease, particularly pulmonary tuberculosis. The medical teams that went in with the advancing forces had special units therefore to deal with tuberculosis, especially among young adults, who were quickly assembled, put through a strict examination, and treated in hospital if necessary.

"What is your age, Christina?"

"Fifteen years, Sir."

Yes, she had lost weight but there hadn't been much food, Sir. Yes, she'd had a cough for the last two years, like some of the other girls, but there was very little fresh air because of the black-outs. No, she didn't cough up much of anything — just coughed and coughed. Pain? Yes, sometimes she had pain right through to her shoulderblade on the left side. And every night, Sir, even in Winter, and even when there were no air-raids to frighten us, we'd be hot and sweating all night.

The nurse got Christina ready for the doctor's examination, and copied his comments on her medical record card.

Appearance? . . . Looks underfed. Wasted. Hair . . . dull. Skin . . . no bloom or freshness — too shiny — white, of course, no sunlight. The blue veins are too conspicuous — may be just lack of cutaneous fat. Few hard glands along her neck — bad sign.

"Now take deep breaths, please Christina." Oh! — restricted movement left shoulder.

"Again please." The spaces about her left collar-bone flat and rigid too.

Steady breathing emphasised the difference in the extent of movements of her chest — the left side seemed lazy and slow; the breath sounds were weak and almost inaudible there and when tapped, there was a kind of dull thud and not the clear resonant note the doctor wanted to hear. Her pulse was rapid, her breathing somewhat too fast and the X-rays showed suspicious changes in the substance of the lung. When, at the pathological laboratory, they put a specimen of her sputum under the microscope, it was found to contain the waxy, curved, acid-fast, four micron long rods of tuberculosis which had taken hold of Christina in the hungry years of war, when the crowded sunless houses had harboured the dangerous germs and when her resistance was low, the sympathetic physician told her.

Resistance! That was a word she knew well. A whispered and ominous word. A word to make men stand erect and proud among their fellows or to crouch in fear at the midnight assault on their doors. She had known men dragged away to torture and death for Resistance. Resistance derails the troop train and blows up the ammunition dump. Resistance helps the prisoner to escape and hides the Jew. Resistance is a force which strikes blows at a hated enemy. "My father belonged to Resistance, Sir", she said proudly.

No, not that kind of resistance, little war-taught girl. These are forces of survival and not destruction and they protect friend and foe alike. They are cells and chemical substances circulating in your blood and composing the tissues of some of your organs, and they can unite with and neutralise dangerous cells which might damage your body.

They are the messages coded into the genes of your chromosomes to perpetuate the production of these chemical fractions and to give you protection all your life. They come from good nourishing food and the fresh air and sunlight.

When the darkness of military domination fell on your country, the tuberculosis bacilli flourished for they love darkness and hate light, especially sunlight. In your crowded air-less rooms, closed to retain such warmth as there was when fuel, clothing and food were luxuries, you were daily exposed to them, sprayed forth by the coughing of your infected fellows till you inhaled them with your breath or swallowed them with your food. Penetrating the moist tissues of your throat or the absorbing surfaces of your intestines, they reached those glands which are part of your inmost defences, and where, in better days their journey might have ended by their destruction. Now, instead, their growing numbers caused these glands to swell to visible size like those now seen along your thin pale neck and those deep in your chest, whose irritating presence caused your persistent cough. Their increase is slow. In the bacterial world where a common reproductive cycle may take twenty minutes, their generation time is nearer to thirty hours. Their waxy water-resisting coat gave them protection against the white cells of the blood, which repeatedly attacked groups of them until the whole system of attackers and attacked grew big enough to be visible as a small greyish fibrous lump about one millimetre in diameter called a tubercle.

As it grew still bigger, the centre, isolated from oxygen and nourishment, decayed till it formed a small pus-filled cavity. If the armies of liberation had not come, the numerous tubercles would have fused in time to form cavities in your lungs. The arteries bridging across these cavities, now without support, may have ruptured, and you could have coughed up your heart's blood in frothy death. But if no such sudden end overtook you, you could have died slowly over the distressful years, your lungs fibrosed, compressed, adherent, distorted, cavitated, your body fevered and sweating from the putrid decay within your chest, or any other part the slow unremitting process chose to attack.

Like others, Christina, you were forced to live in the crowded dark, cold, and hungry slums of war, just when your young growing body at its most susceptible age needed food, warmth, light, fresh air, and sunshine to build itself a constitution to defy the infection, for of all the factors which resist tuberculosis a sound constitution is the most precious.

But what is there that is precious and can survive in man's most evil environment — the environment of the conquered — despised, displaced, entrapped, enslaved, expendable, starved, hated, unwanted, a hindrance to the mighty War Machine when nothing else but it mattered?

But the armies of liberation did come, bringing you food and warmth for your cold starved body, and in due course, streptomycin and ioniazide to heal the disease within you; and freedom.

When she was allowed from hospital, Christina went down to the harbour to see if, through all the misery and tumult of those vicious war years, little Langelina had survived unhurt. There she sat on her granite boulder, still gazing into the water, but Christina fancied she could see a few puzzled wrinkles in the bronze brow as though she were thinking how strange it is that men who could love the beauty of a mermaid's statue, would yet treat a little living child with bestial cruelty.

"And perhaps", thought Christina as she remembered the grim years of war, "for the *second* time in her life, tears moistened the eyes of the little mermaid".

4

SYPHILIS

The spirochaete Treponema pallidum, eight microns long and 0.2 microns in width in the form of a fine spiral of eight coils is the cause of human syphilis. It is difficult to stain with the usual microscopic dyes and is usually examined by the special technique of dark ground illumination. Though it is easily destroyed outside the body, it is resistant to chemotherapy and is very difficult to culture.

When microscopic organisms invade a larger animal they do so for purposes of their own. They have their needs of food and oxygen and warmth like all living things, and if at times they find this conveniently supplied by either man or beast and flourish therein, it is usually with malice towards none! The body they use as a home often can adjust to their presence, and a stable partnership which is usually harmless or even beneficial to the host may result. It is only when the host and its new tenant are completely incompatible that disaster and death may ensue. It is a pity that the study of microorganisms in the past has so often been related only to the study of disease. We should not hate them all because a few have criminal tendencies.

Why then does this spirochaete of syphilis which is found only in man so completely destroy him?

It belongs to a large family of spiral organisms which are found right throughout the world. You can find them in every healthy mouth. You may see them in most pond water. They infect rats, mice, dogs, cattle and swine, lice and ticks and even some shellfish.

But of these multitudes of spiral organisms, none will produce syphilis except Treponema pallidum, the syphilis spirochaete. Nor will it grow and produce syphilis in any animal except man and a few close monkey relations.

Is there something therefore, in human flesh, which it can find nowhere else? Does it recognise a quality we cannot discern, and does this imply that we know less about ourselves than the spirochaete of syphilis knows?

Sobering thought!

Though it may come with no destructive purpose, this slender living spiral corkscrew is carried in the blood to every part of the infected body, and the defensive human cells, apprehensive of foreign proteins, attack it everywhere. In the heart, in the arteries, the bones, the brain — the spirochaete stimulates the body to a fierce antagonism to proteins it realises are not its own.

The result of this internal strife is chronic irritation of vital structures, and the soft body cells are changed into coarse thickened tissues incapable of normal function.

Soft and silent at first, the spirochaete saturates and alters every organ and in these it plays a murderous music. It makes the hardened heart valves beat a funeral march like muffled drums, the heart's blood to spin a humming melody in the narrowed arteries, and murmur and gurgle through the dilated aneurysms.

It changes the sweet regular rhythm of the cardiac cycle to a wild, swinging arrhymthia and to the accompaniment of this harsh orchestra its host will dance his way with jerky shuffling steps through his final years unbalanced in both body and mind.

It seems that the spirochaete did not set out to cause such damage, but that the body in its vicious opposition to a protein its genetic information tells it should not be there, has reacted so violently against it as to evoke fibrous changes in its own tissues.

Maybe in a few more million years the body and the spirochaete may learn to tolerate each other better. Certainly they are bitter adversaries just now.

BABETTE LA DOUCE

"The fathers have eaten sour grapes and the children's teeth are set on edge."

Ezekiel 18;2

There was nothing remarkable about Babette at school. She and her friend Yvonne may have been a trifle more daring than others, a little more attractive.

If the gentle Sisters of their Convent school exerted mild discipline, their rich parents indulged them a little more to compensate. It is unfortunate that over-indulgence sometimes leads to idleness, idleness perhaps to reproof; reproof to resentment.

A pity that Babette's sweet voice, patiently trained in singing by Sister Mary Theresa, should sometimes scold and quarrel with those who loved and cared for her. And that as they grew older both Babette and Yvonne should have seen their cultured homes and respectable families as prisons of domesticity, conforming and old-fashioned.

Well, Paris is a big City. They would find other homes!!

If you pass along the little street which runs downhill to the main entrance of Gare St. Lazarre in Paris on any fine evening, you will see young women standing about carelessly and with little apparent interest in the local scene. But if you observe carefully you will see that they are as alert as any other hunter, for this is their hunting ground; their vineyard, their rendezvous. Here they earn their living, shared usually by a male sponsor. A precarious living. With a predictable and unenviable end.

Ask not how they came to be here. They travelled many roads. A little youthful waywardness perhaps; deceitful wine; frustrated love; wild company; sexual precocity, or so often just the tempting alternative to plain bleak poverty, — accommodation, food, wine and payment are tempting baits.

So here they are, diverse in background and personality, yet with a common aim. Nothing egalitarian in their appearance, nor in their speech and actions. They are all their own shop windows, and the merchandise is displayed as attractively as possible. They smile, they hum a little song, they tap the pavement with a dainty foot; they promenade with aristocratic grace.

Amongst them tonight are Babette and Yvonne who should be sitting in their own homes with maids to wait upon them. But the path which leads from this little sloping street back to their parental firesides is wet and slippery with tears. It was enlivened often with alcohol and the tinkle of wineglasses. To this place it led Babette and Yvonne five sad years ago.

Babette had been the pampered darling of a rich and fastidious man, but was not wise enough to put some of his bounty away. There were rivals. The hold of a mistress is a tenuous one, not to be risked by any meanness of dress or personal adornment, and fashion is expensive. Youth went by and the bloom began to fade. The rivals were younger now and more attractive, and Babette found that it was not only within the concept of evolution that the fittest survived.

Well — there are more men and fresh liaisons. Why break your heart over one? As so another lover — and another — and another.

The lovers are less generous now, less fastidious; and her fading mansion costs more to keep it from falling into cosmetic ruin.

So Babette stands here in the twilight near the entrance to Gare St. Lazarre drumming with her foot and humming her little song. Sing on Babette as the swans do. Besides being hostess tonight to whatever

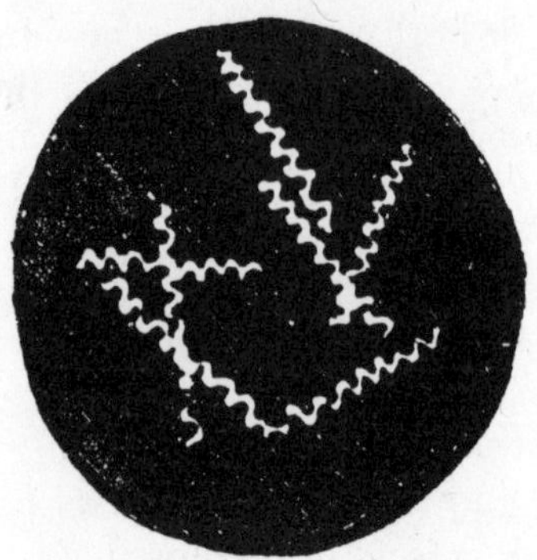

Treponema pallidum.

strolling males your little song attracts, you are now hostess to a little twisting spirochaete called Treponema pallidum. He has come to you perhaps from far across the ocean, perhaps from Naples or London or Hamburg or Marseilles or maybe he is native to your own Paris. But he is yours now; and unlike your many lovers, he will never leave you. If you have a baby, (which perhaps your lover will disclaim), he will share it with you. You and he, your children, your future lovers, and their children, are now just one big family.

He is not big but he is handsome. He is about eight times the one thousandth part of a millimetre long and one five thousandth part of a millimetre across, or more simply eight microns by 0.2 microns. His shining body forms a delicate spiral with eight neat turns one micron apart. He twists and bends and he expands and contracts his coils. At times he even forms a shining circle of silver, but only for a moment before he springs back to be a dainty, erect spiral again. He is truly a lithe and lovely gymnast, a dancer of international fame. He rejects the tawdry colours of the microscopist's dyes, for, like any ballerina, his stage must be dignified and subdued. He is best seen then on a ''dark-field'' where the light does not come up through the condenser of the microscope to illuminate him, but is transmitted at a tangent across it while the viewing field remains black. Then he is seen at his silver best. A spotlighted beauty, gyrating against a background of black velvet, flexing, twisting, coiling. He is visible under the microscope by directly transmitted light like other organisms, only if he is stained with silver nitrate. Then he appears black, motionless and degenerate on a yellow stage. Here we see the villain unmasked; for gay though his dance may be, his beauty is no deeper than his radiant reflecting surface. His character is blacker than the dark-field he dances across.

When he first found you Babette, he stayed for a time where he was, dividing into two about once a day to establish a family, and the little sypyilitic sore he formed there, was merely where he planted his standard to denote possession. He is a traveller really — a piratical, destructive traveller, who will journey through all your organs, even to the very structure and marrow of your bones on which his evil marks may show even years after your death.

Like your lovers, Babette, he will adore those red lips, that snowy skin and that witty tongue and he will leave on them the imprints of his fervent kisses; and he will so possess your loving heart that he will live right inside it on its very valves, and in its most intimate arteries.

Into those regions in your brain which beget thought, reason, and imagination, he will plunder his calamitous way, till at last you will exchange your bodily misery for hallucinatory happiness and travel untroubled and uncomprehending towards an imbecile grave.

At first Babette had small inconvenience with her gay and social little parasite. A few sores perhaps, and ulcers on her tongue and lips; but better a sore mouth than an empty stomach, for to ask for treatment would invite restrictions. Some tender glands, a rash, some fever, a swollen knee, nothing really serious, and then she was well again. So thought Babette! She even married and set out to have a family. Now the forgotten spirochaete, attracted by the newly developing embryonic cells within her, passed from her circulation to the blood of the unborn child. She miscarried. The next time her baby was still-born. And the next. Despondent but determined, Babette carried another. Great joy! It was born alive. But the little spiral fellow in her blood was born alive with it, so that it died before she had time to love it. Then at last her daughter came after anxious months. She lived, but was not as pretty as Babette had been. Her cry was hoarse. Her nose was depressed in the middle so that she had constant ''snuffles''. She was deaf, jaundiced, and wizened; her nails fell off and her hair fell out, and in her mouth were ulcers the same as Babette herself had had so many years ago.

When the babe was a month old it was covered by a red blotchy rash and they told Babette at the Clinic that it was filled with the descendants of those same spirochaetes which came to her in her youth from Naples or Hamburg. It would not be the support and comfort of her old age. If it lived at all, it would probably be subnormal and would need all her care instead.

The Clinic could have told her more. It could have told her that though he was now harder to find than he used to be the faithful Treponema was still with her and could be detected by the ravages in her blood. That he was deep within the linings of some of her biggest arteries. That their once strong elastic walls were bulging dangerously

---

FOOTNOTE:

This is the usual sequence of births to a syphilitic mother.

because years of spirachaetal irritation had made them thin and weak. That her heart valves could not close firmly because they were irregular and hard from the same cause.

That her bones were hard as ivory in some places but soft and yielding in others, so that her leg bones would soon begin to arch forward with her weight. That the marrow in them was poisoned and could not manufacture the blood cells on which her life depended. That she would ultimately lose her memory, and intellect and become unfit even to care for the infant whose maladies had brought her to the Clinic. That she may at last become something unintelligent, incontinent, salivating, revolting, — an animal idiot thing in the tattered remnant of human form, devastated over the unhappy years by only eight microns of twisted silver spirochaete.

Snow drifted quietly before an icy stinging wind — a wind in which no old lady should be out-of-doors. A bitter wind which could bring the chill of death to old bones, especially when the little flesh upon them is diseased, abused, aged, ill-fed and poisoned by the toxins of vicious germs for most of a lifetime. When the old heart that once raced with delight to the embraces of forgotten lovers, now races only to compensate for the inefficiency of its beat, and is embraced only in the painful grip of anoxic disease; when the arteries which once brought blushes to pretty cheeks, and decked rosy lips to receive stolen kisses, are clogged and narrowed by syphilitic irritation and their thin walls dilated to bursting from the same cause.

Old Babette and her spiral companions of the years faced it bravely. She was no longer Babette of the streets — she was Babette, rich, powerful and beautiful, a friend of all the great. She took her unsteady way along the Seine bank towards the iron bridge which crosses the river to Notre Dame where she would pray before she went home.

Notre Dame Cathedral, Paris.

Some boys playing in the piling snow shouted ugly things at her, but with the shreds of a once real dignity she walked resolutely on and ignored their rudeness when others of her sisters would have sworn and shouted back. As she neared the little coffee shop which faces across the bridge, Madame la restaurateur saw her dimly through the steamy windows and called her in.

"Babette you should not be out tonight. I shall give you some hot soup and a little wine to warm you and you shall stay with us until the snow has ceased."

The faded grey eyes looked up gratefully from the faded grey face partly hidden under a faded grey bonnet weighted down with snow, as she accepted her friend's hospitality; and then, warmed by the wine and the company she played the piano for them and sang as Sister Mary Theresa had taught her in that distant country of her youth where there was no want and no wickedness and respectability was stuffy and conservative.

It was too late then to go to the Cathedral, so she plodded home along her dismal street. Lately she had found it hard to ascend that long stair to her room, and the gnawing, grinding pain in her chest had been getting worse. Though elated by the wine she could not climb them all tonight so she sat down half way to the top, shook the snow from her shawl and wrapped it round her cold legs and the pain in her chest grew worse than ever. At last she struggled to the top. It was then that the gnawing pain became a tearing agony plucking out her heart, and with a feeble shriek she fell unconscious.

"The Gendamerie have asked if you will need a port-mortem on the old lady who died suddenly in the Rue d'Amour last night Sir."

The Senior Physician looked up from his desk.

"Who was she, Doctor?"

"We knew her only as Babette, Sir."

"Oh, Babette? She was a plus four syphilitic with aortic incompetence and an aneurysm which I suppose ruptured last night. She was an aristocratic old lady and a great friend in her youth of Yvonne in the G.P.I. ward, who will probably drag out her demented senility for years there. Babette is luckier.

No. I won't need a post-mortem, doctor; the well advanced syphilitic shows so many signs during life that we rarely need any further examination after death. Just fill out the certificate please and I will sign it."

5

BOTULISM

The dangerous microorganism of Botulism lives in the soil and though little known by the average person it produces the most deadly of all the forms of food poisoning and is particularly hazardous because it does not produce the symptoms popularly associated with food poisoning. Food which can be fatal may appear to be perfectly wholesome when eaten. The spores this organism produces are highly resistant to heat and can persist in imperfectly prepared stored foods. The toxins growing from them are destroyed by moderate heating however so that suspected food can be rendered safe in this way.

In the pickled beetroot on the pantry shelf, or the preserved beans or maybe in the fish paste or the sausage meat, the cook may have hidden the most potent of all poisons.

But though the cook has handed you something more deadly than an asp, do not blame her. The cunning seeds of botulism, present in the soil of every continent and more prevalent in virgin than in cultivated land, which indicates an occupancy longer than man himself have been uprooted from their normal job of harmlessly changing Earth wastes back to fertile soil again. Taken to the kitchen with the beet or the beans, they have been boiled long enough to kill the adult organisms but not their resistant spores. If the cook had boiled them for something like twenty hours or heated them under steam pressure at 120 degrees Centigrade for half an hour all their spores would have been killed. Now they are placed in sealed tins or jars and stored on the pantry shelf.

Like all living things they must have oxygen, but unlike most they will not accept it from the air but must obtain it by breaking down other chemical substances. So storage in the full airtight jars is exactly what they want and they proceed to grow in this suitable environment

Clostridium botulinum showing spores.

and to manufacture their most remarkable toxin which has neither taste, smell or colour but only an unbelievable toxicity.

The most malevolent poisoner of the Middle Ages could not have concealed his murderous purpose more cunningly than the innocent housewife, for the food when opened will have no taint, or perhaps the mere suggestion of it and most probably it will be eaten with relish. And once eaten there is practically no antidote.

In a day or less, the diners will show signs of a fatal intoxication and if much of the offending food has been eaten or there is a delay in giving antitoxin there is little hope of survival.

In her defence, let it be said that even food never seen by a cook may have similar fatal results.

When wild waterfowl show botulism in large numbers the epidemiologist must seek an unusual solution. Have they eaten the carrion of the contaminated waters, the airless sludge of the swamps, or do the botulism-carrying fish, in some manner, pass it on?

More than fifty thousand cattle die of it each year, mainly because the mineral deficiences of much grazing land leads them to chew the bones of any skeletons they may find — things such as rabbits or frogs, toads, lizards, snakes or birds. The decaying flesh of the dead animal would offer those airless conditions in which the botulism spores will grow into bacilli manufacturing their toxin.

So this son of the soil, seven microscopic microns long, useful in the vital bacterial job of re-cycling our nitrogen and fertilising the earth, living harmlessly in the intestines of animals and mankind and ensuring its survival in difficult times by forming heat-resistant spores, has another side to its character. Put it where it has nutrients but no oxygen, and it can manufacture the most potent toxin known to Science, and one which has the singular ability to pass unaltered through the intestinal wall, and the subsequent power of destroying that chemical which transmits nerve messages to muscle fibres, so bringing bodily activity first to a partial condition of paralysis then to a general stop. It can kill an ox or a man with equal ease.

Grow it suitably in broth culture free from oxygen and one litre of this innocent looking broth will kill nearly half a million guinea pigs. Purify the toxin to a crystal form, and a microgramme can kill more than twenty million ordinary twenty gramme white mice if there were found such numbers of victims or so heartless an executioner. It is the most fantastically poisonous substance known and even this dreadful toxicity can be increased by laboratory manipulation.

Fortunately, it produces an antitoxin and so can be effectively prevented in cattle, but human cases are so rare because of the perfection of modern canning techniques and the comparatively rare use today of the home canned product that immunisation would be unnecessary.

"WHO DUN IT?"

The hotel which also operates a general store and post office is the centre of activities in many small country towns especially in America and Australia. At such an establishment in a mid-western American town, two travellers arrived one afternoon in the fall of 1936 to find there the proprietor and his wife and one other boarder, who seemed to know who the strangers were though there was none of the friendliness to be expected between former acquaintances meeting by chance in this distant place.

These travellers decided to stay for a few days at this quiet little hotel and in doing so they also decided the time and manner of their deaths. No violence overtook them. They simply ate their way to certain destruction.

Some weeks previously, the landlady had purchased vegetables from an itinerant farmer for sale in her store. When some of the string beans amongst these became wilted and unsaleable, she canned them herself rather than let them go to waste, using a process now rejected as unsafe, which had been developed and recommended by Governments during the First World War as a means of storing food and known as the "cold-pack" method.

Some days after the arrival of the two new guests, the household sat down to a mid-day meal of boiled bologna sausage with potatoes, bread, butter and coffee. At the special request of the regular boarder, who went to get them himself, some of her canned string beans were added to the meal. To the landlady's dismay on opening the jar these beans seemed to have a musty odour and looked a little foamy as though they had been recently shaken up well. The newcomers assured her they would be fine if rinsed in cold water, and so she did this at the kitchen sink five times. The regular boarder still declared that were rancid and refused them but the landlady and her husband and the two new guests could now find nothing unpleasant about them and ate them with relish, teasing the abstainer about his fastidious tastes.

Within forty-two hours there were four corpses in the little hotel. Only the fifth diner at this tragic meal — the fastidious one — survived quite unaffected.

The little town was stunned. The popular hotel keeper and his wife and guests had been poisoned at their own table! In their midst an unidentified poisoner! Though everyone justly suspected the regular boarder, he strangely made no attempt to flee from the scene of so obvious a crime.

At the Coroners Court, formal evidence was given of the arrival of the guests, their unusual decision to stay for a few days in this rather dull little town, their apparent former acquaintance with the regular boarder, and the lack of the expected friendliness between them. The regular boarder was a somewhat introverted quiet man, a school teacher, taking classes in chemistry at the local school with full access to poisons and a knowledge of their properties. He was by no means popular and never spoke of his past from which it was commonly supposed he was a fugitive.

Evidence of the hotel housemaid who worked only in the mornings and who had left for her own home as soon as the family was seated at lunch, showed that the regular boarder was the one who suggested using the canned beans, that he had gone to the pantry to get them himself and that they appeared to have been made a little frothy, presumably by being violently shaken up to mix some added ingredient. Though it was his own special request that they be added to the meal he refused to eat them, even after the rinsing in cold water and the removal of the frothiness had made them appear quite normal to the other four people. It was the landlady's regular custom to can her own fruit and vegetables and no one had ever suffered injury before.

The suspect himself, the regular boarder, gave evidence, but refused to disclose his past to the scrutiny of a small rural community which was already quite convinced of his guilt. He admitted that he had known the identity of the two unfortunate guests, and though he did not know their present occupations, they had not been policemen when he had known them.

FOOTNOTE:
The deaths of these four persons did actually occur as described in September 1936.

He was not a cooperative or willing witness and did nothing to dispel the very strong suspicion that he had deliberately poisoned these two gentlemen who were investigating him on behalf of the Law on some serious past crime, or who had perhaps been his accomplices whose silence could be assured only by death.

The doctor who had attended the four deceased persons then described their signs and symptoms which he was certain were due to poisoning, though his diagnosis of the particular poison was still awaiting confirmation by the laboratory in Chicago to which he had submitted specimens.

When questioned about the possibility of foul play, he would say only that all four had died in the same way within forty-two hours of the meal on September 14th, that in his opinion they died of poison present in the jar of string beans served at the meal, and that the fifth person had been unaffected because he had eaten none of the beans. He would not attempt to explain why food that appeared good and edible to four persons to whom it was offered should be rejected completely by a fifth person at whose special request it had been provided.

The Court therefore adjourned to await the laboratory report and reassembled in one week. The week had been filled with rumours all of which, in the form of confession, arrest, suicide, or flight had confirmed the general suspicion that the regular boarder was the murderer. It was a tense Court that awaited to hear what potent poison he had maliciously placed in that innocent looking jar of beans to cause four sudden deaths, and presumably lead on to a trial for multiple murder.

After the opening formalities, the Court asked the doctor if he were now ready to give a precise diagnosis of the cause of the deaths of the four persons following a meal on September 14th. The doctor had received the laboratory confirmation of his diagnosis and was directed to inform the Court what, in his opinion, this poison was, and how it could have been present in a part of the food at that meal.

"The poison was botulinum toxin, and it had been generated in the jar of string beans inadequately processed by the landlady herself in their preparation" he announced quietly.

The tension in the Court snapped like a steel wire, and an astonished chatter followed. Though the Court itself was somewhat unwilling at first to accept the diagnosis, the findings of the laboratory were irrefutable — "Botulinum toxin Type A in lethal quantities". All the remaining jars of beans had been impounded and submitted to examination also. All contained the fatal toxin.

Nothing remained then but to bring in a verdict of death by misadventure but before closing the Court, the doctor was asked if their could be any means to prevent such tragedies in the future.

Yes, the doctor said, there were means of prevention. The organisms of botulism like those of tetanus to which it is related, are found everywhere in the soil. In thousands of soil samples taken from all over America, including glaciers and high parts of the Rocky Mountains, it had been found in all but one. It is therefore always a risk where vegetables or fruits contaminated even slightly by soil are being prepared for storage. It forms heat-resistant spores which can survive the common preserving processes, develop into active botulism germs, and produce their fatal toxin in the sealed jar. Stale material should never be used nor should the once recommended ''cold pack'' method. A pressure cooker having an accurate thermometer and gauge and holding small amounts only to ensure uniform heat penetration should be used for the full prescribed time at the full prescribed temperature. No canned food with a disagreeable odour or showing gas should ever be eaten but if this is necessary, it can be made safe by actual boiling for fifteen minutes, because though the spores are extremely resistant to heat, the toxin once formed is easily destroyed by it. All suspected foods should be burned or buried.

The doctor finally suggested that press and radio programmes and schools should undertake to educate people to the very real danger of Botulism and its comparatively easy prevention. There was no television to assist this publicity in 1936.

With thanks to the doctor, the Court rose and the little mid-western town settled back to its usual rural calm, feeling perhaps a little disappointed that so promising a sensation had died away so quietly.

6

INFLUENZA

Tiny and versatile — a mere one hundred millimicrons or so in diameter — the little 'flue virus exists in different types, and even changes these from time to time. Though so small, it is quite resistant and can exist on such articles as blankets for weeks. When freeze-dried in the laboratory it can be kept alive indefinitely. No wonder it is such an active little fellow.

In 1918 armies from around the world were returning to their home countries after the first Great War. But many of the men who had lived through the dreadful battles of a world at war did not survive the first few months of peace. They fell ingloriously to the attack of an invisible virus before they ever saw home again, and the virus, travelling home with others, killed there, no fewer than twenty-five million people. And then, as though satisfied at having taught the quarrelsome human race to know that it alone had no monopoly of slaughter, the little virus disappeared, and hasn't been seen since. At least, if it has returned it hasn't been seen in such a murderous mood. When it disappeared it left behind some difficult questions which no one can answer today, for laboratory techniques in 1918 were not good enough to identify the villain or to retain any evidence which might incriminate him by later investigation. Today he would be arrested, identified, convicted, imprisoned and finger-printed for we could now grow him on fertilised eggs, we could identify him by a series of agglutination techniques, we could convict him by measuring the antibodies present in the patients blood and their neutralising effect on laboratory cultures of the accused, we could imprison him in freeze-dried cultures where he would remain alive for years, and we could finger-print him by storing antibodies from his victims for future comparisons with antibodies produced in other victims by other possible suspects, if the crime were repeated. Today we would know what criminals we were dealing with; in 1918 we didn't know.

What biological change produced so dangerous a killer from the little-loved but still moderately law-abiding virus of influenza?

Such a pandemic arising so suddenly and fading away so completely can make us appreciate the explanation our ancestors would have supplied, that it was a divine visitation to punish war-like men for the wickedness of global war. The heliotrope discoloration and purple blotches on the victims would have reminded them of the Black Death, and it would have seemed the most acceptable story. Can we find an alternative?

First maybe it wasn't the 'flu virus at all but a new disease. However, it was diagnosed by many competent doctors around the world as showing all the symptoms and signs of influenza, albeit in an exaggerated form, and we must accept their diagnosis. If it were a new disease, where has it gone?

A mutant — a new form arising from the parent virus by a new arrangement of genes? Mutants often persist to become new species and this one had such wide incidence in every country of the world, that surely some would have become established somewhere.

Was it only our ancient acquaintance the flu virus, with temporarily increased virulence which again reverted to its old ways. If so, what caused this increase and can it happen again? It was said that this increased virulence was due to the massing of many people together in unusually dense communities which gave rise to an unusual degree of virulence by frequent passage from person to person. This is known to enhance virulence, but why didn't it happen in the next World War when greater numbers were engaged? It has been suggested also that it was not the 'flu virus which killed, but its blood-thirsty companions, the pneumococcus or the staphylococcus. Why then this sudden fatal partnership when they have been frequent past companions anyway, and if *they* have changed and not the virus, what happened to *their* suddenly acquired extra virulence?

The fact is we don't know. Neither do we know if such an event can be repeated. Although we have learnt much about microscopic organisms since 1918 we can never afford to underestimate their power. We can just be grateful for the essential part they play in Nature and hope that we can continue to learn.

Here is what we do know about this extraordinary virus.

It is spherical, between 80 and 120 millimicrons in diameter and therefore invisible, except by the electron microscope. It is of three types and several sub-types or strains with a constant tendency to change which makes its clinical recognition and its immunological control very difficult. At intervals of about 20 years there seems to be an abrupt change in the virus giving virtually a new one and producing world-wide pandemics. Perhaps the 1918 pandemic was one of these which somehow got out of hand. But that was 60 years ago and we haven't seen it since.

It can be grown in one of the fluid cavities of the developing chick embryo or in a broth containing monkey or human embryo kidney tissue.

It does not seem to be confined to humans for strains have been identified from pigs, horses, ducks, chickens and a variety of wild birds. Useful research on it really dates from the fortunate discovery that the ferret is also susceptible and can be used as an experimental animal instead of humans, but the use of fertilised hens' eggs for the culture of influenza virus has now completely superseded the ferret. Protection by immunising vaccines is not totally effective but is being improved.

THE EGG AND YOU

The eminent virologist, gave a few ominous sneezes and had a shivering attack or two, and by the time his secretary was able to persuade him that he might have "the flu" and would be better at home in bed instead of in a rather bleak laboratory, he had also a severe headache and aching limbs, a high temperature, a dry mouth and throat, a severe cough and painful rheumy eyes.

Next day he was in bed with his pains extending even to his aching eyeballs, and his scalp feeling that someone had tried to pull all his hair out by the roots, still reluctant to admit that the virus of influenza on which he had worked for years, should have exactly the same effect on him, an acknowledged authority, as it might have on any farmer's boy who wouldn't know a virus from a turnip.

A week later, exhausted by the sweating fever and the continous cough, too weak to go back to the exacting routine of a research laboratory, he had to admit that he'd really had a good dose of his own virus and had better go for a rest at his brother's farm in Devon with its gentle hills, its patchwork fields, mossy stone fences and green hedges.

Around the fireside one evening when he was recovering, the family asked Uncle Charles to tell them something about his work with viruses, and he invited them to put their questions. "What would you like to know first, Evelyn?" he asked. "I'd like to know", said Evelyn, a pretty seventeen and therefore free to take any liberties, "how anyone like you Uncles Charles, could be silly enough to spend a whole lifetime looking for something too small to be seen, and who knows they exist anyway?"

"You can't see electricity Evelyn, but you can recognise its effects. Its just the same with viruses. Their effects were first noted on the farm you know and I'd like to tell you about viruses on the farm — about eggs, cows, ferrets, ducks, potatoes, cheese, milk, bees, pigs, silkworms, rats, mice, rabbits, horses, sheep, dogs and even tobacco and dahlias.'' They had never thought that a virologist in a laboratory in London could even know about such common things.

He described the discovery of the larger microscopic life like bacteria and then told them about the extracts made from tobacco plants infected with mosaic disease, which although microscopically free from all visible bacteria, were still able to transmit mosaic disease to healthy plants. How foot-and-mouth disease of cattle was then studied in the same way and carried from sick beast to healthy beast by filtered extracts which were shown to contain no visible microscopic life, so that it was obvious then that there must be invisible agents of disease in these infective filtrates. They were therefore given the name of ''Filterable viruses''.

Having proved their existence, the next thing was to measure and study them.

''Uncle Charles'', Evelyn gasped, ''you surely are not going to tell us now that you can measure the size of these invisible things! How could you possibly measure something you can't see?''

''By laboratory techniques, Evelyn. We use graded filtering membranes and find the lowest one which, because the filtrate is infectious just allows the virus to pass through. Or we spin a suspension of them in a high-speed centrifuge, much as you use a cream separator in your dairy, and measure the rate at which they sediment. The influenza virus must be spun at 40,000 times the force of gravity for thirty minutes to make it sediment.

We can measure the size of the larger forms of microscopic life by fitting an ocular micrometer, which is a fine glass disc with lines ruled

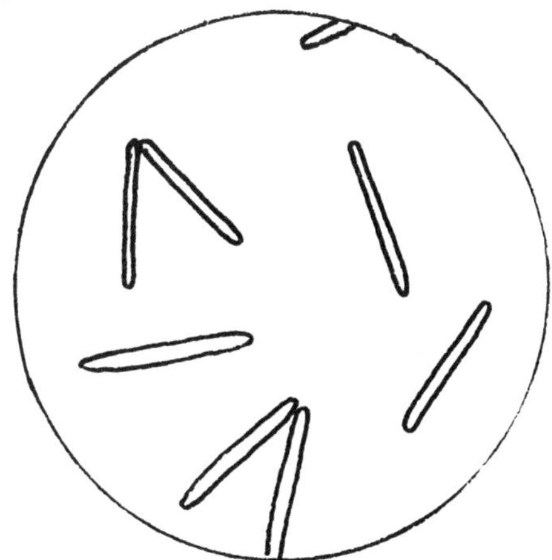

Tobacco Mosaic Virus seen by the Electron microscope.

on it, into the eyepiece of the microscope, and calibrating it against a stage micrometer, which is really a tiny glass ruler used in microscopic work, and which is fitted on the stage of the microscope where we will then place the organisms to be examined. To see them better we use aniline dyes to stain them especially the Gram method.

"But Uncle, if a thing is invisible, how would you describe its size? Would it mean anything to say that Virus X is an invisible part of one inch long?"

"We use millimetres, not inches in scientific work Evelyn. Then we divide the millimetre into one thousand parts and call each a micron, and again we divide the micron into one thousand parts and call each a milli-micron. It is of course impossible to conceive such a small measurement but now with the electron miscroscope we can have pictures of them, even if we cannot see them. The smallest viruses are about 10 milli-microns and the largest about 300 milli-microns in diameter."

He took from his pocket some electron microscope photographs of the virus of smallpox looking like small cubes of sugar and said their size was 200 x 300 milli-microns.

When Evelyn asked if he knew what they were made of, he replied that the purified virus had been shown to contain protein, fat, carbohydrate, copper and two vitamins.

"Well, Charles", his brother said, passing back the photographs, "It's incredible that we can have photographs of something invisible isn't it?"

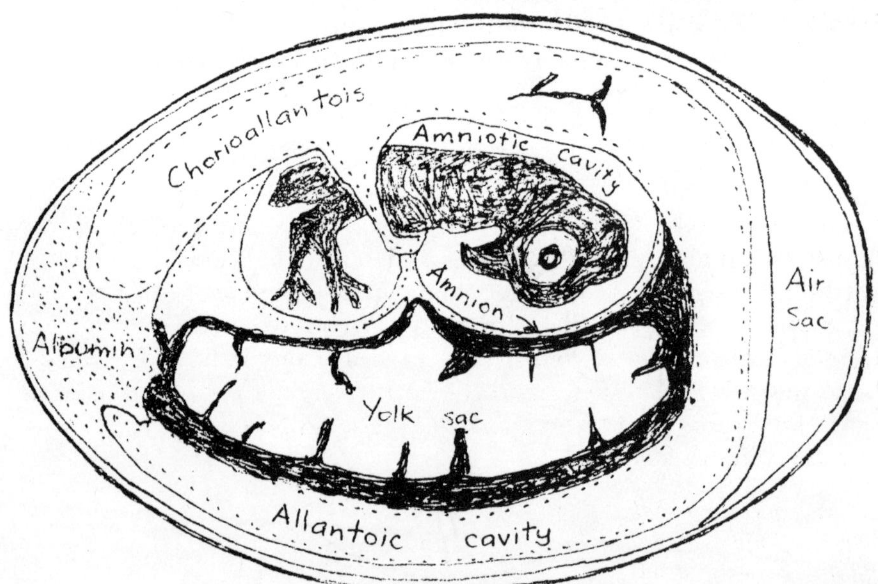

Fluid chambers in the developing fertilised egg.

"Only invisible to us Tom. But there is another world besides the one in which we move. A virus, similar to the one you see there, probably exists in millions of millions on the udders of your milking cows. The infection it causes is called cow-pox, and because it was found to protect milkers against smallpox, it was the first organism to be used in the process of protection which has now become the highly involved science of immunology."

"The virus of influenza which brought me to Devon is smaller than this pox virus, only about 110 milli-microns in diameter and has an internal core of 70 milli-microns."

"Now the invisible thing is being divided into its invisible parts", said Evelyn, slyly winking at her brother, who also thought that Uncle Charles was really only amusing them.

"At first we found it impossible to grow influenza virus in the laboratory."

"To grow it, Uncle! Who would want to grow the flu virus?"

"We grow it to study, and to make vaccines but we thought it would grow only in human beings until 1933 when we found that it also grows in ferrets."

Evelyn convulsed with laughter and the others joined her at the thought of a ferret with the 'flu and carrying a box of tissues.

"And what were you going to tell us about eggs, Uncle Charles?"

"I was going to tell you that we grow viruses on eggs."

The laughter exploded again.

"We never thought you were such a wag, Uncle Charles."

Now what does an eminent virologist say to that? It could make him realise that the common routine techniques of the laboratory so scientifically simple to him, might appear like black witchcraft to the lay mind.

"Round about the cauldron go,
In the poisoned entrails throw."

His laboratory dealt in rats and cats, dogs and rabbits, toads, hamsters, guinea pigs, mosquitoes and bugs; the blood and sera of sheep, horses, men, goats, rabbits and deer; fertilised hens eggs and pregnant mares urine; the kidney cells of monkeys, of rats and of human beings; sputum and urine; seaweed, sour milk and potatoes; bone marrow, the testicles and the corneas of rabbits, the spleens of mice, the brains of rats and monkeys, muscles of cats and guinea pigs, and the venom of snakes.

"Toad that under coldest stone
Days and nights hath thirty-one
Sweltered venom sleeping got,
Boil thou first in the charmed pot."

What a gruesome array!

And it had never seemed curious to him before.

Infusions, broths and cultures.

Tinctures from plants good and evil, the chrysanthemum, or deadly nightshade, or poison ivy. Pollens, acids, toadstools, histamine. Sinister dyes of green, purple, black and orange —

His shelves were full of them!

Dropsical fluid, hydrocoele fluid, amniotic fluid, cerebrospinal fluid, digested proteins, bacteria, viruses, moulds, algae, fungi and the latest addition — the saliva of cattle ticks.

"Fillet of a fenny snake
In the cauldron boil and bake.
Eye of newt and toe of frog
Wool of bat and tongue of dog
Adders fork and blind worms sting
Lizards leg and owlets wing."

His thoughts flew back to London where they awaited his return.

His familiars! Cats, monkeys, ferrets, bats, frogs, rabbits, rats and toads.

His charms! Enzymes, pepsin, papain, ox gall, dehydrated liver, yeasts, malts, whey, and poisonous cyanides.

"Liver of blaspheming Jew
Gall of goat and slips of yew
Slivered in the moons eclipse,
Nose of Turk and Tarters lips."

He wondered if Shakespeare had ever worked in a research laboratory!

"Scale of dragon — tooth of wolf.
Witches mummy . . ."

He remembered that in some of the latest work even human embryonic and foetal tissues were being used.

"Finger of birth — strangled babe
Ditch — delivered . . ."

He was recalled by Evelyn's twinkling eyes and her merry laughter. "Go on Uncle Charles, tell us more. I'm only a farmer's daughter."

So they thought he was just romancing!

Then Charles explained how a lucky chance after many years investigation, showed that only the ferret of all animals can be infected by throat washing from human cases of influenza applied inside their nostrils, thus giving a ready source of the virus for laboratory purposes, and replacing human volunteers, and how finally a better method was found by inoculating the virus into the various chambers of fluid which develop within the growing embryo in the fertilised hen or duck egg.

"And so it's really true about the ferrets and eggs, Uncle?"

"Absolutely. Viruses are totally parasitic and will grow only on living tissue. A developing fertilised egg contains suitable living, growing, sterile young cells — so today hundreds of thousands of fertile eggs are used for virus culture in laboratories around the world. There are about 10 millilitres of allantoic fluid in each egg and each millilitre will produce about 10^{10} (ten to the tenth power) of infectious units of virus which equals ten thousand million — 10,000,000,000 for each egg. In this way the influenza needles you get each year are prepared — "The egg and you, you see."

"Any why, Charles", his brother asked "are these 'flu needles not as reliable in their effects as some others?"

"Because Tom, plants and animals and viruses differ in their properties and we have to go along with Nature. Some organisms have such an effect on the body cells which form antibodies to them that this ability lasts for life, others leave their antibody forming impression on these cells for a short time only. Influenza virus is one of the short-term kind. Measles virus leaves its impression for life.

There is another reason. Most viruses enter the circulation and so make close contact with the antibody-producing cells; influenza virus is inhaled directly to the lungs and avoids them.

And finally, there are many strains of influenza virus, and vaccines prepared from one strain are not totally effective against others."

Evelyn, still rather suspicious now wanted to know how a thing like cheese could possibly get sick.

"Because the bacteria used in the starter cultures to make the cheese, are eaten by their ultra-microscopic enemies Evelyn and so the batch of milk just remains uncurdled and flavourless."

"Big fleas have little fleas, eh, Uncle?"

"Yes, exactly. That old tag of Swifts once used to discredit the idea of microscopic life, is all too true. Microscopic bacteria do have ultra-microscopic viruses which destroy them. These are called 'bacteriophages' indicating their power to digest other organisms and we know quite a lot about their chemistry and how they do it."

"And does your work have much bearing on farm life, Charles?"

"Yes. Viruses are everywhere, Tom. The garden dahlia, for example is an important carrier of the virus of spotted-wilt disease of tomatoes and it is spread from one plant to another by juice sucking insects exactly as diseases like yellow fever are carried between humans by bloodsucking mosquitoes. And potatoes, Tom — you grow them here, don't you?"

"Yes, Charles, and we get new seed each year from Scotland or we'd have very poor crops."

"Because Tom, in Scotland, the insects that carry virus diseases to potatoes are not very active, so that each year you make a fresh start and avoid building up a concentration of viruses which would reduce your crop."

"What filthy pests these viruses are Charles!"

"Not exactly, Tom. In some parts of the world, the potato moth is a most troublesome pest destroying both tuber and plant, and we have been able to control it by using a virus which kills the larvae of the moth. So that here the virus is on the side of the farmer. It is possible that such biological control with viruses will in time replace all those unpleasant insecticides."

They all appreciated these scientific explanations of accepted farming practice, and wanted to know about bees and silkworms and dogs and mice and rabbits, and all the others.

It was a happy and interested family that made ready for bed at an unusually late hour after Charles had told them of the sacbrood of bees, the rabies which destroys dogs in Europe and infects vampire bats in South America, the virus and other diseases of silkworms which fired the original bacteriologists to save an expiring commercial industry, about the myxomatosis which was used to produce fatal epidemics in rabbits in Australia, because they had increased to fantastic pest proportions; of the virus diseases of fruit and plants, and then of measles, mumps, poliomyelitis, yellow fever and the 50 or so viruses that can afflict humans, of distemper in dogs, of swine fever, cattle foot and mouth disease, and of cancer and its possible relation to viruses.

He told them about recent studies on viruses and forecast that man may know what goes on in outer space before he is able to say what goes on in the ultra-microscopic body of a virus.

With the thought that the minute architecture and physiology of viral cells might hold more mystery than the boundless Universe, the family retired, grateful to Uncle Charles for giving them a peep at knowledge whose existence they had never suspected.

7

PNEUMONIA

Though the inflamed condition of the lung either scattered throughout or consolidated in one lobe which is called pneumonia, can be caused by the inhalation of irritants, it is more frequently due to the pneumococci, pointed organisms enveloped in pairs in a capsule. It may be divided into seventy-five different types by serological means. Commonly found in the throat they appear to induce disease only in the presence of some factor which either increases their virulence or which depresses the resistance of the host.

Pneumonia used to be called "the old people's friend". The aged and the infirm who now often stay unhappily on when the time for departure is long past, imprisoned in a frail body which has accomplished all that it can accomplish except to withdraw itself from the tedious contest, were once gently carried off by the welcome pneumococcus. With the minimum of pain and distress, the little fellow consolidated the lobes of the tired lungs, the oxygen reaching the blood steadily failed, the brain became clouded and the patients' life ended quietly only perhaps a few unwanted years before that inevitable event.

But doctors intent on keeping alive the spark of life began to use bottled oxygen to aid the failing lungs and the pneumococcus suffered its first defeat. Then it was "typed" and serum used against it. Then the sulphonamides. Then penicillin. Then other antibiotics until today, though still about in noses and throats, it is little feared unless the patient has been severely exposed to cold, irritating gases, alcoholic intoxication, malnutrition, or some viruses.

Even in the absence of all Man's antibiotics and sera the pneumococcus is not without enemies. The protective white cells of the blood — the phagocytes — are their natural foes and attempt to eat them; but they have learned to grow a resistant capsule around themselves. The phagocytes cannot digest this outer coat and would

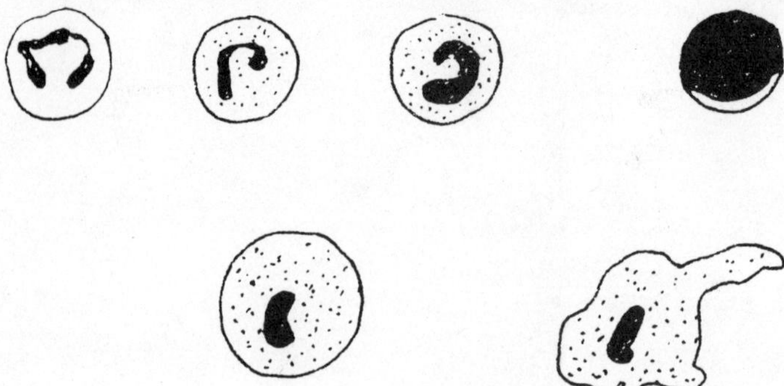

Phagocytic and antibody-producing white blood cells.

now be quite ineffective had not some of the patient's body cells also learned a lesson in biological adaptability. These special cells now manufacture a substance — call it an antibody — to coat over the invaders coat and when this is applied, the previously ineffective phagocytes at once digest and destroy the foreign pneumococci. These cloak and dagger tactics, which take about ten days, result, if successful, in the clearing of all pneumococci from the blood and the patient recovers.

What manner of germs then, are these, that going into dangerous places, always go about in pairs and wear protective clothing. And how do these utterly remarkable organisms change their chemical coat of armour against the white blood cells so readily.

It would be a well-stocked shop that could offer no less then seventy-five kinds of the one coat, but that is the choice offered to the pneumococci. There are seventy-five different types of pneumococci and the differences lie in the chemical structure of their capsular coats to which the protectice body cells can, with equal skills, produce seventy-five different antibodies.

Is it not likely that such organisms, trying to survive by seventy-five different means against seventy-five different antibodies, might be able to find a way to defeat man's antibiotics as other organisms have done? Penicillin, especially, has destroyed them rapidly since it has been used against pneumonia. How long can we expect the pneumococci to accept defeat?

And when shall Man accept the truth that a few new laboratory techniques and the accidental discovery of a few antibiotics do not give him mastery over the microscopic world which he is hoping to understand; nor explain such bewildering mysteries as the occurrence of seventy-five chemically different structures in so insignificant an article in Nature as the capsular coat of a usually harmless microscopic inhabitant of human throats.

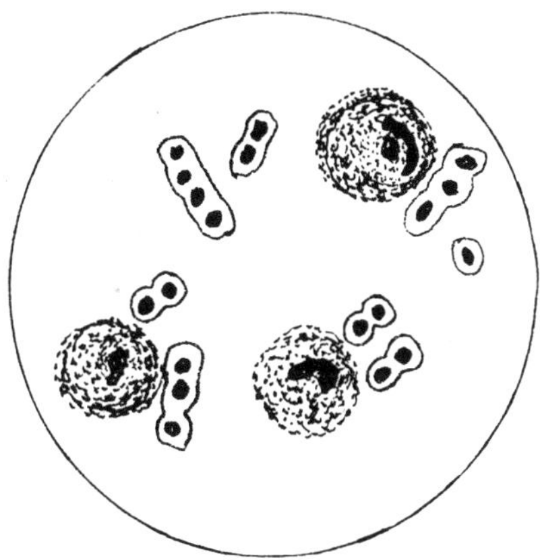

Klebsiella pneumonia.

SEVENTY-FIVE VARIETIES

It was a bright sunny morning in Colombo, but from time to time showers blew in from the Indian Ocean, squalled across the harbour blotting out the white-painted ships, pattered down on the sandy beaches, raced over the grass and up the coconut palms, scurried across street after street of the scattered city, looked in at the window panes of the lighthouse, passed over the shops and markets of the native quarter, the fashionable suburbs and the beautiful Cinnamon Gardens, leaving the bright tropical plants there washed and glistening in the sunshine which always pursued these racing showers, chasing them in from the ocean, over harbour, town, and market, to make them flee for their lives up to the hot and steamy hills. You would think the little showers would really give it all up and leave the place to the sun, but they never did. Just as soon as one shower was driven off and all was sunny again, another would rush in only to be driven off too.

The smart sentry in his starchy white uniform with shining brass buttons who kept guard at the entrance of Government House near Queen Victoria Gardens would no sooner come out of his dry sentry box to parade up and down looking so soliderly, his gleaming bayonet sparkling in the sun each time he wheeled about at the end of his beat, than down would pelt the showers again, and, all his military might being quite useless against such a foe, he would retreat to cover in his sentry box once more.

The beggar on the sidewalk shook the water from his crumpled rags, stretched the crooked leg on whose pathetic deformity he depended for a living, and settled back again with patience, for in a life of poverty who would care about a little rain.

The humble oxen plodded on with their awkward carts, patient, laborious, and obedient, shower or sunshine. While it rains they shake their big ears to toss away the raindrops and when it is sunny to drive away the flies.

The banana sellers put up their umbrellas and stay at their posts. A little rain would never hurt their wares, hanging in bunches from the iron railings of the gardens, or standing in heaps on the pavement against stone walls.

On the harbour, the white painted launches carry passengers from the liners to the pier, splitting the blue water in two; the dark athletic crews as handsome as their craft, straight and dignified as their polished boat hooks; and just as expert in their work with the sun shining on smooth water or the showers and squalls tossing their boat with delight.

The traders on their sturdy dhows go about the business of getting them ready for sea. Moored side by side, their high black prows and furled brown sails look disdainfully down on the sophisticated launches. The dhows share life with the winds and the sea. They are ancient aristocrats. The trees which gave them hulls and masts, and spars and rudders were watered by the showers and warmed by the sunshine of forgotten centuries. The propelling winds which fill their lanteen sails have carried them wherever they had wished to go from remotest antiquity, and would drive them far on into the future while they have masters who know and love the challenge of the sea.

Boys, untroubled by the demands of school are everywhere, following the tourists or swimming and diving about the ships to recover coins thrown into the water as the cool evening breezes dispel the remaining warmth of the hot tropical day.

Here, on the pier with his friends, sat little Jaysera, bare to the waist, enjoying the coolness of the rain and the warmth of the sunshine.

In his throat, as in the throats of half the people in the world, the organisms of pneumonia lived harmlessly. Today however, something upset this delicate balance between host and parasite, so that as Jaysera went homewards in the dusk with his tired mates, the usually harmless germs took on a virulent quality and became stealthy killers.

His temperature rose and Jaysera, with hot burning skin, felt the cold chills of infectious disease and shivered through the long unhappy night. His fevered blood lit his eyes with the dark fire of black opals, and at last made his brain a twisted labyrinth of delirium. He strained to leave his bed and swim with the other boys around the great white ship in the harbour, to look up at those tourist faces at the rail, clear for a moment and then hazy again as though he saw them from several fathoms deep.

The doctor was called. He noted the high temperature, the dry cracked white-coated tongue, the sores on the lips, the flushed cheeks, the bright eyes, the hot dry skin, the dilated nostrils, the increased pulse and breathing rates, the short painful grunt as the troubled boy breathed, the thick blood-stained sputum; the position of greatest ease lying on the affected lung to keep it still. He applied his stethoscope. The inflow of air was much diminished. He tapped the chest wall and the sound over the left lung was dull and heavy as though he had tapped on a barrel full of fluid, although on the right side it continued to give out the resonant vibrant sound of emptiness.

Then he told the mother that her son had left-sided lobar pneumonia and he feared he might die from the failure of his lungs to supply the minimum needs of oxygen.

As the disease progressed, the white cells in the boy's blood, which usually engulf and destroy foreign materials, called up reinforcements from other tissues and increased in numbers from the normal seven thousand to sixty thousand in each millilitre, and they fiercely attacked the pneumococci. At first they were unable to harm them enveloped in their resistant coat, but the body, with a knowledge and accuracy we cannot explain, recognised the exact type of pneumonia organisms from the seventy-five possible variants and prepared the exact antibody against them. This newly-made substance, by some chemical or physical attraction, coated itself round the protected foreign organisms and neutralised their defences. Triumphantly then the white blood cells flung their amoeboid arms around them and rapidly digested them all.

With his body cleaned of the dangerous attackers, the boy's temperature fell rapidly to normal, the heavy consolidated lung became light and aerated again allowing painless breathing and the doctor knew his little patient would soon be back smiling at the tourists again.

8

DIPHTHERIA

The organisms of Diphtheria are neat rod shaped bacilli only six microns long but possessing a remarkable power in their minute invisible bodies to make a venomous substance which irritates the tissues of the throat till the swelling and tough adherent membranes forming therein can strangle a patient, and which so inflames the heart muscles that death becomes a race between suffocation and heart failure. It is heart failure that usually decides the unhappy contest.

It is distressing to walk through an acute diphtheria ward where the healthy children of a few days ago lie dying from the poisonous toxins of a preventible disease, but it is now fortunately, difficult to find such a ward, for immunisation against this dangerous organism is now almost universal, and only because of occasional parental carelessness is the doctor sometimes confronted by a case of diphtheria in developed countries, though it can still be seen amongst the less fortunate.

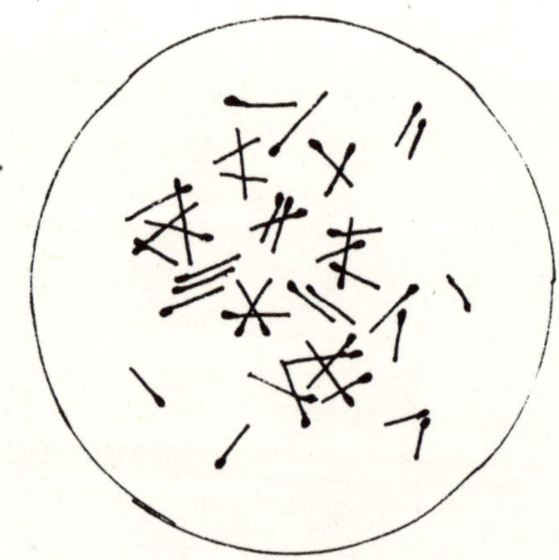

Corynebacterium diphtheria.

It is a century and a half since the disease was first recognised, and almost a century since Klebs described the bacillus which causes it; and for the last half century we have had safe and efficient means of preventing it, yet many little ones have died of diphtheria, with health and safety no further away than the nearest immunising clinic.

Though this little bacillus has been such a monstrous murderer it has now confessed to the microbiologist the whole story of its own crime, the means by which it prepares its deadly toxins, and ways to neutralise them and prevent the slaughter of future generations of children.

Not that it was easily persuaded to turn informer!

The knowledge has been wrung from it over many years of patient questioning by many skilful interrogators in the bacteriological laboratories of many countries.

It was given the specific name of diphtheria after Loeffler showed that it was always present in the membranes in the throat of diphtheria patients. This is really a tall poppy amongst bacteria, being a straight slender rod, often with club-shaped ends and a tendency towards branching which is more characteristic of the fungi. When, like other bacteria, it multiplies by simple division, the two parts may remain partly attached so that an appearance of Chinese lettering is built up as its numbers grow. He is a fastidious fellow, and if you want to grow him in the laboratory he must have ideal conditions — temperatures of just about human body heat; good food — he likes a nice beef or sheep serum broth with plenty of glucose, some of the vitamin B group, no acid, and just the right amount of iron, and plenty of free oxygen. In such a broth he can make diphtheria toxin and in the throat of a child which has never been immunised he will manufacture this same deadly poison which is absorbed into the child's blood. The bacillus itself remains in the throat and corrodes the flesh, causing it to form a protective membrane which may grow big enough to strangle the child whose heart and kidneys are meanwhile poisoned by the toxins circulating in its blood.

There are cells in its body though which can manufacture antitoxin, a chemical capable of neutralising the diphtheria toxin. If enough of this antitoxin is produced quickly enough, the child will live; if not it will die. The remarkable thing is that these body cells can recognise diphtheria toxin from any other toxin, always produce the appropriate antitoxin, and retain this ability for life so that the recovered patient never suffers from diphtheria again.

Now, if we could place this diphtheria toxin in the blood of a similar child in quantities insufficient to kill it, would this same process occur? Experience and laboratory examination show that essentially it does, not only in humans but also in horses which after such injection have antibodies to diphtheria toxin in their blood. Here, then, is a ready source of diphtheria antitoxin to be harvested and stored for use on the ailing diphtheria patient.

And here too is a safe and extremely efficient way of giving every child the chance to have antitoxin in its blood in readiness for a possible future bout with the strangling germ of diphtheria. This process is called immunisation. It is used in the prevention of a number of diseases in man and animals and is really only an artificial use of a process that goes on continually in Nature.

THE MANCUNIANS*

From forge, and farm, and mine, and bench,
Deck, altar, outpost lone,
Mill, school, battalion, counter, trench,
Rail, senate, sheepfold, throne—

Creations cry goes up on high
from age to cheated age;
Send us the men who do the work
For which they draw the wage.

The wage slaves
Kipling

Betty Beaucroft was born in the Midlands, in the city of Manchester, in the suburb of Ardwick, in one of those many terraced houses which were built at the time of the Industrial Revolution as England changed from an agricultural to an industrial nation, and it was found profitable to house the workmen and their families as close as possible to those dark Satanic mills. It made it so much easier for the "knocker up", who clattered down each street an hour before dawn, beating on the doors so that all hands might be inside the mill gates before 6 a.m. to start the twelve hour day. That was of course after 1843, when Sir Robert Peel was Prime Minister and had passed a Bill to limit the hours of work of "young persons" to twelve hours a day. Before that, it might have been fourteen or sixteen hours. It was better, too, that these "young persons" who were aged from five to six years upwards, did not have too far to walk to and from the mills in the cold dark foggy Lancashire mornings and evenings or they might fall asleep exhausted on the streets and in doorways.

The Beaucrofts had lived in Manchester since the trade depression following the Napoleonic wars, when it was impossible to find work in

---

*MANCUNIANS — Manchester was formed on the site of the ancient Roman fort of Mancunium and its present residents proudly use the name for themselves.

the rural areas, and when the power of steam promised to provide plenty of jobs in these new factories in cities with such modern luxuries as gaslight. Waterloo had banished Napoleon to St. Helena. Now, England, first away in the industrial revolution, would have peace and prosperity with all Europe at its commercial feet.

But the common men who had marched together against Napoleon soon found that they were not allowed the right of combination against the capitalists of the new order. They might advance at Waterloo flanked by the splendid cavalry units, but when they marched to listen to an orator in St. Peters Fields in Manchester five years later, the cavalry charged straight at them and cut them to pieces. Waterloo was one thing, Peterloo was another.

Family tradition had it that Betty's great-great-great-grandfather was there at that sunny Sunday morning massacre at Peterloo, and that his revolutionary spirit had come down through more than a century of Beaucrofts. He had fifteen children, for when adults were refused work at nine shillings a week their children might be employed at three shillings, and he who had a large family might eat while others starved. That was one reason why the population of England and Wales rose from nine millions at the beginning of the 19th century to double that number forty years later. Many of his off-spring had never reached adulthood to earn the full nine weekly shillings though. Fog, grime, rain, cold, starvation, tuberculosis, utter exhaustion and beatings to keep awake during the inhuman working hours, are no friends to children.

But that was long ago and things were better now. Betty's house stood right against the pavement of the bare front street in line with all the other houses and there were no spaces between them. Each house was exactly the same as its neighbour. Each entrance had an iron railing. Each railing had eighteen bars and three steps led down from the left to a cobbled pavement.

At the back of each house was a small yard paved with brick and surrounded by a brick wall surmounted by a gabled concrete top acting as a watershed, so that exactly one half of all the rain that fell, and beat, and washed, and drizzled down on it in that wet Lancashire climate ran into Betty's yard and one half into each neighbour's. Each moss-grown wall had an opening with a gate leading to a narrow, dirty, dark, back alley.

But in Betty's yard some Beaucroft ancestor had planted a pear tree which every Autumn cast off its leaves to make a squelchy mass with the rain, and every Winter defied the dull sunless days with its long bare arms sometimes weighted down with snow. Then in Spring its pink waxy buds opened, and suddenly it was green with leaves and white with blossoms. Except for this pear tree there was nothing to brighten the frowsy sameness of the frowsy street. Her ancestor of Peterloo would have approved of that pear tree. It lifted men's eyes from the dead pavements for a while and tuned their ears from the silent monotony of brick and slate to hear the lively music of a living tree.

When the Council Medical Officer came to the school, to immunise the children against diphtheria, Betty's parents, guided by the dread of innovation, and hostility to the things they did not understand, decided that Betty should not be immunised.

Great-great-great-grandfather Beaucroft of Peterloo would have raised a dissenting voice. Was not Peterloo all about innovation? Wasn't it their social enemies that wanted to keep things as they were? The old gentleman would have taken Betty's parents to the nearest infectious diseases hospital — there is one at Salford — to see for themselves the tragic, trusting eyes of dying children . . .

Come into this diphtheria ward, Walter and Beatrice Beaucroft; there you may learn what diphtheria is. Here is Susan, just Betty's age — would have been a fine woman. Why are her eyes so tired; her skin so muddy-grey, her breathing laboured?

It is sad about Susan. She will die tomorrow or the next day. When she was brought to hospital, the toxin of diphtheria had so disrupted the muscle fibres of her heart that even antitoxin could do nothing to save her. She will sink away quietly as the little poisoned heart slowly dies.

Some of these other childish faces watching us anxiously from the cots could be out of the world in another week.

There is a disturbance at the door of the ward and anxious attendants hurry in with a little blue-faced child; a mother in tears;

poor suffocating child with the ugly membrane of diphtheria cutting off her breath.

Only one thing can save her — tracheotomy. Expose the wind pipe, cut into it and part the rings of cartilage. Insert a metal tracheotomy tube. Now air is sucked in and the little blue-black face turns to livid purple, the purple fades to blue, the blue to dusky red; a little pink shows about the lips, the ears, the nose.

Now the bewildered eyes open and close and open and roll and close again in terror. The little, nearly strangled body writhes and twitches, the heaving ribs fight for the oxygen of life and consciousness slowly returns with its breath.

A nurse will stay day and night beside that child's cot, ready with a sterile swab to clean out that life-saving tracheotomy tube every time a plug of mucus or a piece of the detaching membrane closes it up to start that terrifying experience all over again. She will keep the oxygen ready, check the fluttering pulse and the irregular temperature, soothe the child's distress as she injects large painful doses of antitoxin and daily examine that horrible membrane now shrivelling and detaching itself. Here Walter and Beatrice Beaucroft, you may gain some knowledge of the horrors of diphtheria and you might be grateful to the ghost of your wise ancestor from Peterloo who brought you here to learn the truth; for he knew that some are always fighting against apathy and ignorance, and the sharp swords of the charging cavalry, to make life better and safer for their fellow men.

They took Betty that summer to crowded Blackpool to see the illuminations. The next week there were deaths in Blackpool and Betty had a sore throat and croup. Croup! The word struck chill to the heart of Beatrice Beaucroft. To her mother's generation it meant death. Her elder sister had died of croup. They didn't call it diphtheria then. The older doctors called it "Morbus suffocans". Betty's throat swab was positive for diphtheria and the toxin circulating in her blood was destroying the cells of her heart, her kidneys, her liver and the fibres of her nerve trunks; her fair, clear skin now a dark leaden hue from slow asphyxia, her ribs and breast bone deeply recessed in the long fight for breath, the automatic breathing centre in her brain struggling to resist the over-powering toxin, Betty was marked to follow Susan and the other unprotected children in the diphtheria ward to the grave.

Those neutralising substances which living bodies, including the human body, can manufacture against harmful toxins, can now be produced artificially in the laboratory for the protection of man and animals. In this work, man's ancient friend, the horse, has become his modern scientific partner. By injecting graduated doses of diphtheria toxin, the horse's body cells can be made to produce antitoxin to neutralise it.

At Speke, not far from Betty's home, there was a Serum Institute and the doctors used these antitoxins in the hospital and on Betty Beaucroft.

So Betty is at school again, saved by the patient studies of men she will never hear of, and possessing now an immunity to diphtheria acquired through pain and anxiety, risk, suffering, expense, and a journey to the very brink of eternity, which could have been safely acquired by a few injections given by the school medical officer. All the Beaucrofts right back to Peterloo and beyond wouldn't have been so lucky. It is for Betty's generation to live in the security of knowledge gained in such institutes and pathological laboratories. If you asked her now what she thinks about fever hospitals, she would reply in her broad Lancashire accent, "joost smashin!"

9

POLIOMYELITIS

How does an infinitely small organism only thirty milli-microns in diameter and therefore not visible even under the microscope come to inhabit the human bowel and exist nowhere else? How did it evolve at all? How did it learn the pathways to the motor nuclei of the spinal cord where it sometimes provokes inflammation which may result in extreme paralysis.

Did human beings themselves originate and hatch this little fellow traveller which can sometimes tragically harm its host or did it arise in some other focus which it permanently exchanged for the alimentary tract of man? Or does it exist unrecognised in other places or in other forms? Though we may have learned much about it, its origins, as with other forms of life are its own private mystery.

Until a few years ago, each returning summer brought with it a real and constant fear, the fear of parents for the safety of their adolescent children, for with the summer came an epidemic of what was then known as infantile paralysis. It came so quickly and gave so few warnings! The healthy child of yesterday, could be in bed today with a mild fever and a few muscle pains and tomorrow could be a case of paralysis, to survive crippled of limb, or not to survive at all. The child had swallowed the virus of poliomyelitis, one of the smallest of all viruses with a diameter of less than thirty millimicrons, and this tiny invader had multiplied in its bowel and set out by unknown routes to reach the spinal cord and the roots of some of those nerve fibres which carry motor messages from the central nerve apparatus to the muscles, and direct their action. Why it should select these vital spots in the whole wide body is its secret alone. So small that it is not visible even with our most powerful microscopes — only a large molecule consisting of perhaps a few mobile chromosomes — yet it knows where it wants to go and it gets there. It knows a man from every other animal except a few monkeys for it will grow in no others. Whatever ancestor of the forgotten aeons taught it to know, and to do, taught the lessons well for it never errs. Why does this infinitesimal speck of living matter take so long a journey from bowel to spinal cord or brain to make its home with such unfortunate precision? We do not know the answer.

Landsteiner discovered the virus in 1908, and it is now known to be of three types. It is fairly resistant and can remain alive in water or sewerage for four months. Though it is noticed only in epidemics, antibodies to it are found to be almost universally present in human beings so that it must be a widespread and common infection with few of the cases showing any clinical signs of illness. Depending on the position and the extent of damage to the nerve roots it may cause only minor indisposition of a non-paralytic type, or paralytic poliomyelitis of greater or lesser degree or even respiratory arrest and death in a few cases.

Fortunately, it readily produces antibodies, and immunity can now be produced artificially by growing the virus in tissue cultures of monkey kidney, and by the preparation from these cultures of vaccines of either the killed SALK type to be given by hypodermic needle, or of the live SABIN type to be given by mouth.

Both vaccines give satisfactory protection, but it is a pity that the healthy fear which once drove people to be immunised has been replaced by a false sense of security because poliomyelitis is now less frequent, and to protect against it is now sometimes neglected in the mistaken belief that it is no longer a threat.

WHAT MYSTERY DISEASE?

"The word 'crippled' to me suggested a condition that could be applied to some people but not to myself. But, since I so often heard people refer to me as crippled, I was forced to concede that I must fit this description, yet retained a conviction that though being crippled was obviously a distressing state for some people, with me it didn't matter.

The crippled child is not conscious of the handicap implied by his useless legs. They are often inconvenient or annoying, but he is confident that they will never prevent him doing what he wants to do or being whatever he wishes to be. If he considers them a handicap, it is because he has been told that they are.

Suffering because of being crippled is not for you in your childhood; it is reserved for those men and women who look at you."

I can jump puddles
Alan Marshall

There was a long-distance personal call for G. Blandford Jackson, already over-busy in his executive suite, with the morning's crowding Company affairs, and he frowned with annoyance at the unwelcome interruption. Couldn't anyone else take the call? Didn't his secretary know he wasn't in?

"It's from Sybil, Sir", she told him.

"Sybil? Oh well, thanks, then put her through, Fay."

Sybil had the weekend free from school and would be home tomorrow night — with some pals, of course. Could Mom arrange a little party? And you'll meet the eight-thirty train, Pop? Sure, of course she had the most wonderful parents in America.

The crowded train arrived on time. The weekend was fun with dancing, singing, tennis, swimming and cycling and next night of course, there was another dance.

"Go to bed? Oh no, Pop, we can go to bed at school. They won't let us play records there like this."

As they waited next day for the train to go back to school they were so tired they just wanted to sleep all the way home.

On the drive back home, the Jacksons agreed that it hadn't been a relaxing weekend for anyone. Wasn't it just amazing where the kids got all their energy? Never still!

The long distance operator had another personal call to G. Blandford Jackson, and this time it was not Sybil's gay young voice at the other end. It was the quiet and serious voice of Miss Crawford, the headmistress.

"Nothing wrong with my kid, is there, Miss Crawford?"

There was.

Sybil had returned from the weekend at home, quite exhausted and Miss Crawford had kept her in bed for a few days. This morning she had been feverish and complained of some muscle pains and a little stiffness in her neck and back. Dr. Boyle had come and had ordered that she be put into isolation. The diagnosis? Yes, Miss Crawford was sorry to say it is poliomyelitis but the doctor says there is no serious nerve damage yet. Certainly she would tell Sybil her parents are coming right away.

G. Blandford Jackson's fiercely competitive business world broke up into a thousand unimportant pieces. The dollar-filled contracts and commercial anxieties of an hour ago were profitless dust and ashes. Poliomyelitis! His kid! The happy active kid that danced and swam and sang, now to face a crippled life on crutches! In an iron lung perhaps! Their Sybil! And the doctors didn't know a thing about it! Even their president had been a poliomyelitis victim.

In deep despair, he and his wife motored all day to the school and waited anxiously while Miss Crawford telephoned Dr. Boyle.

Dr. Boyle had completed his work for the day but would see them at once — at his home.

Terrified of what they would hear, they rang the doctor's door bell. The doctor took them into his comfortable lounge and asked his wife to bring in some coffee before he noticed the anguish in their faces. Mrs. Jackson, fighting back her tears, asked if he thought Sybil had a chance of recovery or if she would have to face life in a respirator.

Surprised by the question and their distress, he told them Sybil was only a mild case of poliomyelitis and he expected her to recover completely or with only the slightest disability. It was most unfortunate if they had not already been told that.

The Jacksons had been so sure of the magnitude of their calamity that for a moment they could not believe the doctor's words, and nervously Jackson told him they had never heard of mild cases of poliomyelitis — only of those heart-wringing histories of crippling paralysis and automatic respirator treatment, life in a wheel chair and all that sort of thing.

''There are bad cases of course and they can be pathetic when they occur in young persons whose life is all before them, but these are really exceptional'', the doctor said.

''It sure surprises us to hear that, Dr. Boyle'', Jackson said. ''But why do medical folk still know nothing about it?''

''Now you folks have both had a bad day. Would you like to go to your hotel now and see me in the morning?'' Dr. Boyle sympathetically asked.

The Jacksons, at least partly relieved from what they thought to be an overwhelming sorrow, asked Dr. Boyle to tell them more about this mystery disease. They were sure they'd never sleep if they did go to their hotel.

Over the coffee his wife brought in, the doctor began to tell them that poliomyelitis is no 'mystery disease' at all, though in the popular mind it is looked upon with dread. It was first properly described in England in 1799 but only since 1908 has its cause been known to be a virus.

Jackson interrupted here to ask if it wasn't a ''kind of pious cloak for ignorance'' to say a disease was due to a virus, because he understood viruses were invisible, and so when doctors didn't know the cause of disease they always said it was ''due to a virus''.

''On the contrary, Mr. Jackson, we know a great deal about viruses. This one is an inhabitant of the whole alimentary tract. We know its size and shape. It is twenty-eight millimicrons in diameter, or twenty-eight millionths of a millimetre, and in shape it is roughly cubical. We know its habits. Like most viruses, it is unaffected by freezing but is destroyed by heat, so it survives well in such foods as milk and ice-cream but pasteurisation destroys it. The only animals in which it will grow naturally are man, monkeys and chimpanzees, though now it can be grown artificially in human and monkey tissues in the laboratory. Because there are a number of strains, investigation of it has been difficult and complicated.''

Jackson didn't want to question the doctor's authority on the subject, he said, but he couldn't believe that anyone could measure the size and shape of something that couldn't be seen, or even know that it existed at all.

''You know it exists by its effects, of course, Mr. Jackson. Effects like we see at the moment on your daughter and which can be transmitted to other persons or to other monkeys. You measure its size by laboratory techniques, with centrifuges and membranes, not by a ruler, Mr. Jackson.''

''Oh sure'', Jackson agreed, obviously quite unconvinced, and wondering how his business would progress if he signed contracts he couldn't see. ''Then you said it was cubical in shape. How could you tell *that* from your membranes and your centrifuges?''

''By taking photographs with the electron microscope, my friend. That has confirmed all former estimates.''

Jackson was convinced now. If you could get it down on paper and have a good look at it, then everything was O.K.

''But why, doc, does it attack so many young people today, when they are better cared for now than ever we were?''

''And we didn't get polio'', his wife added.

''You've probably had it several times. In most big American cities today, nearly every child has polio by the time it is four and one in every four of them has it twice by the time it is six. And only the exceptional few show any symptoms at all.''

This was news indeed! That is, if the doctor was being honest with them, but Jackson thought he saw inaccuracies in the argument. In the business world, there was always someone trying to hoodwink you and he didn't suppose these scientific fellows were any different.

''But aren't you supposed to be immune to polio once you have had it, Doc? How can these kids get it twice?''

''I told you there are different strains. A new strain might cause a new attack.''

''I see Doc. You could show symptoms from one strain and later on be infected by a different strain.''

''That's it exactly Mr. Jackson.''

''O.K. then Doc, but don't you tell what people have by their symptoms, so how can you say they've had it if they never show any?''

''By measuring the antibodies in their blood. These can't be present unless the patients have had the infection. Most people do show these antibodies when they are tested for them.''

"Then if nearly everyone has these things in their blood, how do you know they have anything to do with polio?"

"We know they are antibodies to the poliomyelitis virus, because they inhibit the growth of the polio virus in the laboratory and because they protect people from polio, and their concentration in the blood rises in the presence of the virus."

"Then why do we have these epidemics?" G. Blandford Jackson, business executive, was determined to get his contract quite clear, before he signed it.

"We get epidemics because poliomyelitis virus lives only in the human alimentary tract and is not found anywhere else. In primitive communities with poor sanitation, there are plenty of opportunities for its transfer and the whole population becomes immunised. With sewerage systems, there is less exposure and the rate of immunisation falls, leaving a susceptible population mostly of young people."

"So this is one disease that tends to become more prevalent with better conditions which is a reversal of the common rules, isn't it, Dr. Boyle?" Mrs. Jackson asked.

"Quite right, Mrs. Jackson. The Scandinavian countries have been leaders in improved social and economic conditions and their first polio epidemic was described in 1891. In America, we didn't have a major epidemic until 1916. Some of the more advanced African and Asian communities have only recently seen such epidemics and in most backward areas of the world, it hasn't occurred as an epidemic at all yet."

"It still seems fantastic though, Doc, that such a minute invisible thing can cripple and kill human beings when it wants to and often it does no harm at all."

"That may be due to the physical condition of the patient. It is well known that recent severe physical stress is an important factor in the cause and severity of the infection. I've been wondering if Sybil has been overdoing the sport lately."

The parents confessed to the whirlwind weekend, the crowded train and the exhausted young folk and realised that their control should have been a bit stricter.

"And now you ask why such tiny invisible things can kill such large animals as we are. These viruses don't attack *men*, they attack *cells*. Myriads of these virus cells attack myriads of our cells so that the effect is not unlike the clash of opposing armies.

Jackson's first doubts were rising up again. He had personally known the insanity of war and its indiscriminate destruction of everything in its path. How, he enquired, could such an attack result in those unchanging symptoms which are characteristic of poliomyelitis and other diseases if the virus cells just attacked the cells of the patient.

Wouldn't that result in the destruction of different parts and so give different symptoms?

Dr. Boyle could imagine G. Blandford Jackson swiftly examining a doubtful business offer till he had exposed every weak spot in the unprofitable proposition.

"You spoke of polio as a mystery disease, Jackson. This is where we find mystery. You were with a commando unit during the war and you knew how to pass by other inviting military targets so that you could reach your specific job and carry out your particular destruction there, didn't you? That's just what these viruses do too. The polio virus probably first enters the body through the tonsils and the intestines, and then enters the blood, which could carry it to the brain and spinal cord, but it seems very likely that it takes the much more difficult route along the nerve fibres instead. The reason for that may be to avoid defensive blood cells just as your commando units avoided opposing forces. Only within the brain stem and spinal cord, does the sabotage begin. Even here it does not attack nerve cells indiscriminately, but only those particular nerve cells from which 'motor' fibres arise — the fibres which initiate and control movement. It is as though your commando units had orders to destroy a power station and nothing else.

The effect then is to produce a soft paralysis of all those muscle groups under the control of the damaged motor cells. Do you see what I mean by mystery?"

Well, the Jacksons were sure beginning to see where the mystery lay and when they were told that the virus never attacked the areas of the brain which control vision, hearing and the mental processes, for example, so that your polio patient is always bright and intelligent, they saw that this much-feared parasite is comparatively mild and strikes tragically only rarely.

Assured that Sybil would not need radical treatment and having discussed the necessity for preventive inoculation which had proved so thoroughly effective, the Jacksons had some more coffee and went to their hotel, weary, of course, but no longer in dread of the unknown and the mysterious.

Other mysteries to which they had never previously given a second's thought, had been implanted in their comforted minds instead — the mysteries in the world of invisible life. By contrast, Jackson's own bustling business world seemed harsh and vulgar.

As he drowsed off to sleep, G. Blandford Jackson murmured, "Say, darling! Whatever made us think the medics didn't know anything about polio?"

"Well, G.B. honey, I suppose you and I have been kinda mistaking those scarey articles in the popular press for real scientific medical papers, eh?"

"Those people were in a dreadful panic", Dr. Boyle told his wife as he came very late to bed, "the usual thing with polio.

Thought it was still a mystery disease, and it took some time to tell them a little about it. I gave them only the rudiments of course. Jackson thought viruses were invisible things quoted by doctors to explain diseases they didn't understand. I wonder what he would have thought if I had gone into the subject of viral architecture? Do you think he'd have appreciated the fact that the capsids of the virus consist of thirty-two sub-units arranged in the form of a rhombic triacontahedron, or that its protein shell is composed of sixty structural units or a multiple of sixty, and that there is a total of 180 polypeptide chains in each polio virus particle with a molecular weight of 25,000? Or that inside this protective protein shell there is a single molecule of infectious ribonucleic acid with a molecular weight of two million?"

"You should have told him, darling. He might have found some use for it in his business — public relations perhaps."

"Like many other people, my dear, their medical thinking is either hopelessly unscientific or fashionably avante-garde with the latest press items on antibiotics and psychology. The complicated physio-chemical processes of the body are wonders they never suspect. They only see their little ills and want a "wonder drug" to put them right again."

"Yes, darling, and do you see it's two hours past midnight?"

"Sorry dear. When I think of these absorbing subjects in the world of the infinitely small, I suppose I talk too much, do I? O.K. Move over please."

10

RUBELLA
(German Measles)

Human beings may not always be desirable friends but at least in the microscopic world they are the first and only favourites of some. The virus of rubella, the coccus of gonorrhaea, the virus of poliomyelitis, and the spirochaete of syphilis select us from all other living creatures to be their only associates and companions, and however much we reject and abuse them they remain eternally true to us and to no others. There must be something good about us.

The ordinary measles of childhood can sometimes be a severe disease with dangerous complications, but German measles which is the common name for rubella is, in children a mild illness, causing only a low fever, a transient rash and some swollen glands, and after recovery the child is immune for life. But when it attacks a woman in the first three months of pregnancy, it is a serious illness, dreaded because of the damage it may cause to the developing embryo. Babies of such a pregnancy may be stillborn or may be born with defects of the heart, eyes or ears.

Because of the lifelong immunity it produces, it is desirable that young girls should contract rubella and if they do not do so naturally, they can now be immunised artificially so that they will not have rubella in possible future pregnancies, and so place their babies at risk. This may not always be totally effective and any pregnant women should carefully avoid contact with rubella up at least to the fourteenth week, after which there is not much risk, because the organs have then been formed and have only to grow to maturity. It is when the organs are evolving from the earliest buds or groups of cells that a virus can damage them, and rubella virus is the chief villain — a very tiny villain. It is only from 50 to 70 millimicrons in diameter, but it can survive storage at minus seventy degrees centigrade for at least two years. With that sort of resistance it is likely to remain forever with the human race — apparently its only host and friend — for it is found in nothing else.

Immunity from the mother protects the child for the first six months of life.

SEX AND SANITY

"For when once the young heart of a maiden is stolen
The Maiden herself will steal after it soon."

"Moores Irish Melodies"

"She never told her love, but let concealment,
Like a worm i' the bud, feed on her damask cheek."

Twelfth Night

It was winter in Norway. To the old and inactive this meant long cold days spent indoors. It meant log fires, cottage crafts to fill the days, evening gatherings of music, song and stories.

To the young, it meant winter sports. There were many places to skate, and as for skiing, didn't Oslo have the world's biggest ski-jump — the Hohmenkohln — just outside the city?

Arne was not yet a champion on the skis though he dreamed of it. Today he had had the best run yet and with another season or two of practice he imagined himself competing with the world's best.

At the club-house restaurant, he seated himself in a warm corner for a meal before he took the electric train down the mountain to the city, and his mind was filled with thoughts of himself; and the fine ski-run he made today.

Then Olga, tall, fair with blue eyes and cheeks glowing from the cold air walked into the restaurant and seated herself across the table from him. The skiing championships were forgotten. They dined together and at the underground station in the square where they left the train, they arranged a future meeting. From then on, they became constant companions and enjoyed a happy courtship.

They skied together or walked to the top of the long road above Hohmenhohlen to the bronze statue of the engineer who built it. Here they could look down the deep valley and beyond to Oslo fjord — that beautiful fjord which unfortunately had figured in the treachery by which Norway was given over to foreign foes a few years before. They visited the Framhuset wherein the Fram was housed — that ship built here in Oslo which had sailed with Nansen towards the North Pole and with Ronald Amundsen towards the South, further north and further south than any other ship in the world: the Museum which held the

Viking ships recovered from burial mounds of ancient Kings; the folk museums which displayed the crafts, the implements, and the houses of past centuries. They met in the great rectangular park which fills the city centre — where Parliament stands on one hill facing straight across the tree tops to the Royal Palace on the opposite hill, and the University, on the third side of the rectangle, looks across the middle of the park to the Opera House. They went to the House of Parliament with its white pine benches and seats simply upholstered in green, which looked so clean and informal, so Norwegian, and they stood before the world globe on its artistic pedestal which the people of Norway have preserved there as a reminder to new generations that freedom can be lost if vigilance is relaxed, to read the shameful inscription.

"From Adolf Hitler to his dear friend Quisling."

They took the train from Oslo across the mountains to Bergen on its magnificent fjord more beautiful even than the one on which Oslo stood, and the funicular railway to the top of the mountain to view the magnificence of ocean, city, fjord and rugged coast.

Arne and Olga knew the responsibility which rested on them once their courting days were over, to pass on undamaged to future ages, the living cells they had inherited undamaged from the distant past — from the million centuries the inanimate gases needed to become a living molecule and the million centuries it took the living molecule to produce nucleic acid which could reproduce itself and so establish the chain of life from generation to generation. They knew that in each human being, there are billions of cells — microscopic factories seething with life and action, contained within an elastic membrane of the thinness of 100 wave lengths of light perforated by pores so small as to be beyond the resolving powers of even the electron microscope. They knew that in each cell there are 23 pairs of thread-like chromosomes, each containing about one million genes which carry hereditary characters — an incomprehensible number of genes to be divided and distributed in orderly fashion in the processes of reproduction and growth.

With the priceless memory of a happy year of courtship and a dignified wedding, Arne and Olga set up a home on the edge of those cool pine forests above the city where the small spotted deer come out to gambol on summer evenings, and in due course they knew that another, a young life, was journeying there to meet them.

The great, the rich, the highborn and the poor and lowly, all were once a microscopic cell about one hundred and fifty microns in diameter. When Olga was still an unborn female child, there developed deep in the safety of her pelvis, two ovaries, or sex glands covered with

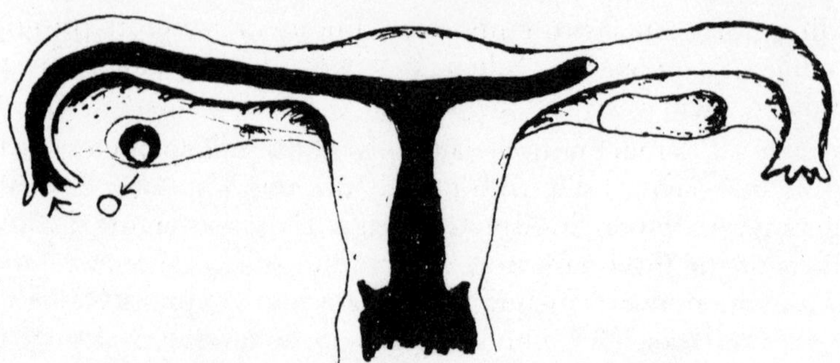

Female reproductive organs showing path of ova at left.

a thin membrane on which lay nearly half a million primary cells which alone of all the millions in her body could reproduce her kind. As she grew into a young woman, these bridging cells, these links between the eternity of the past and eternity of the future, formed successively into mature ova.

Each month the ovary broke the little yellow cyst surrounding one of these and flung it contemptuously off to die in her abdominal cavity unless it quickly found a male fertilising cell to carry it on to new life.

Of all the mighty currents which have carried explorers and merchants across the oceans and along the great rivers of the world, none has had such influence on the destinies of mankind as one which is so small as to be scarcely measurable and so hidden from sight as to be almost unknown. Yet every human being who was ever born has drifted along its gently flowing stream, for it is the current set up in the fluid of the female abdominal cavity by the inward-beating action of the tiny beckoning fingers surrounding the opening of their mothers Fallopian tubes, and it has softly propelled them all from the exposed and hostile site where the ruptured ovarian follicle has cast them, to shelter and safety in this tube. Carried half-way down it towards the uterus, this travelling ovum has met the ascending spermatozoa, and, fusing with it has become the seed of an Emperor or a cobbler, a genius or a fool, a sinner or a saint, or just one of the ordinary millions of mankind. So behind all the human race, its history, its triumphs and its disasters stands a tiny current a centimetre or two long, transporting an expelled ovum from the certainty of destruction in the female abdominal cavity, to a promise of future life in the uterus.

And when Arne too, was an unborn babe, similar glands appeared in his pelvis and then moved down to the external scrotum, for the cells they were to nurture needed a temperature lower than the warm nidus of the pelvis. Within the substance of these testes were myriads of fine tubules, and lining their walls, the same type of membrane which covered Olga's ovaries — that "germinal epithelium" capable of budding forth in astronomical numbers the minute protoplasmic custodians of new life, when in time, the hormones of puberty stimulated them.

Unlike the female ovum waiting in solitary hopefulness, when the time of marriage and fruitfulness had come for Arne and Olga the fertilising cells which he supplied for the propagation of their child was of the order of three hundred million, but only one reached the ovum by travelling upwards through the uterus and along the Fallopian tubes at a speed of about 3 millimetres per minute, and after fusion, the now fertilised cell returned to the uterus to implant itself in the wall and become a living child cushioned round by a sac of fluid and anchored to its mother and her life-giving blood supply by the umbilical cord.

In Olga's body, the child grew to maturity, and at last transferred to the crib waiting in the nursery. And other children followed and made the home happy with their laughter.

And then as if to show how very thin is the shield which protects the delicate tissues in the growing buds of embryonic humanity, tragedy came. Olga fell ill with a fleeting rash and enlarged glands on her head and neck. As she was again six weeks forward in a new pregnancy, she saw her doctor. It was rubella, a disease she had escaped in childhood and had not been immunised against. The doctor explained the possibilities — Olga might have a child as normal as her others, she may miscarry, or, the embryo could suffer damage to its unfolding organs which could leave it deaf, or mute, or with a damaged heart or defective vision. The Gamma-Globulin which he administered may supply some antibody effective against the virus — we could only hope.

But the little one was not so fortunate. This usually benign virus, only sixty millimicrons in diameter inhaled with her breath, pervaded Olga's blood in every part. It entered the placental reservoir of blood which carried oxygen and nutrients from Olga's circulation to her baby's, and diffused through the vessels of the child. It crowded every cell in the developing baby as it had done in the mother, especially the primary buds of some of the organs which were beginning to unfold and destroyed their substance.

''Like a worm i' the bud'' it fed, not ''on her damask cheek'' but on her future heart, ears and eyes, blocking the growth of the primitive cells into those highly specialised tissues with the remarkable ability to see, or to hear, or to form themselves into a muscular pump which would circulate her blood for a life time.

When at last the little one came to join the family, it had neither hearing nor speech nor a normal heart. Innocent little unborn victim of accidental biological antagonism, it seemed to evoke a greater love and found a greater welcome as it needed greater care.

Maybe it is good for us to know that the unfailing accuracy of

Mother Nature in forming the plant from the seed and the flower from the bud, can sometimes err and bring forth something less than perfect.

It may teach us that if the presence in the body of such a benign foreign substance as the virus of rubella — the German measles of our childhood which comes and goes almost unnoticed — can destroy or distort an unborn human life so tragically, other things could not be harmless.

If we understood more of the noxious effects of tobacco, the destructive action of narcotics, and the slow toxicity of alcohol on some of the thousands of millions of microscopic cells and the millions more hereditary genes which the body contains, we might try to treat that delicate living machinery with more respect and greater care.

SCRUB TYPHUS

Scrub Typhus is found in the Eastern parts of the world, as far as Japan in the north, south to parts of Northern Australia, westwards to Thailand, Burma and India, and eastwards to some of the islands of the South West Pacific, but nowhere else on the globe because the mite, a parasite of small bush animals, which carries the rickettsiae causing it is found only there. Like all rickettsiae this one — a rounded fellow 0.5 microns long — is completely parasitic on the living host cell within which it grows, and it is therefore cultivated chiefly in the embryonated hen's egg in which it can produce a concentration of one thousand million per millilitre. In the patient it inhabits the cells lining the small blood vessels, especially those of the heart and brain which swell and decay with their incredible load of parasites. In some areas up to sixty percent of Scrub Typhus patients die.

From the mite to the flea, from the flea to the mouse, from the mouse to the elephant, and from the elephant to the whale, macroscopic animals come in all shapes and sizes, and it is much the same in the microscopic world. The viruses range in size from about ten millimicrons to four hundred millimicrons, and are diverse in shape and structure. Next come the rickettsiae which bridge the gap between the sub-microscopic viruses and the comparatively huge bacteria. Beyond about 0.5 microns, which is near to the upper limit of rickettsial growth, are bacteria, three, four, eight or more microns in length with an occasional larger giant amongst them, and in shape spherical, rod-like, spiral or comma-shaped with a variety of appendages. Larger again are the spirochaetes, the fungi, and then, greatest of all, the protozoa — all made up of one single cell. Animals composed of more than one cell, are classified as Metazoans and include all the larger forms of life.

Though the scientists may label and classify them into Kingdom, Phylum, Tribe and Family, the living things themselves recognise no such artificial boundaries and dwell together. The Protozoan finds a convenient home in the Metazoan. Bacteria inhabit kings and queens and earthworms. Viruses are happy in the borrowed cells of cats or canaries, human beings, dogs, reptiles or vampire bats, and the rickettsiae make their mobile homes in mice and men, cattle, sheep and goats, lice and ticks, fleas and mites. In scrub typhus we have such a union, where rickettsiae living in a jungle mite which sometimes sucks human blood for food, are able to transfer thus to the human being. Such a meeting is purely accidental, as man and mites are not constant companions. Man and his repulsive body louse however, are often resident under the one shirt, an association which has produced many dreadful epidemics of louse-borne typhus. Helped by poverty and misery in overcrowded jails, in packed migrant ships, on pilgrimages, with armies in the field or the refugee hordes which followed them, the rickettsia of louse typhus has killed its millions, until recent years raging incessantly in Europe. It found the Moors fighting the Spaniards at Granada in 1849 and while they fiercely killed 3,000 Spaniards, it quietly killed 17,000. It sat down with the French army besieging Naples in 1528 and destroyed it. It enlisted with the Swedish army in the Thirty Years War and claimed 18,000 victims. It joined the Serbian forces in 1915 to kill 150,000. Just as the First World War ended, it visited Russia and added three million to that country's already grim casualty list.

It has been overseas too. In the hungry years before the middle of last century when the potato crop failed it skipped over to Ireland to stop the hunger of 700,000 with rickettsial death and when the penniless Irish tried to emigrate to Canada it emigrated with them. When in 1847 seventy-five thousand set off to cross the Atlantic, twenty thousand died of typhus and five thousand reached journey's end in a burial at sea. It even looked in on Mexico to kill two million.

It is fortunate for Eastern populations that the particular rickettsia in their midst has rural and not urban preferences. This fellow resembles his European brother in shape, size and habits, but not, luckily, in his choice of a home. His mobile mansion, just visible if your eyesight is good, is a mite which lives on many small jungle animals for which in its newly-hatched larval state it lies in wait on blades of grass. If its mother mite had been infected with rickettsiae, so would it be, for she can transmit this microscopic organism to her offspring in her microscopic eggs. Should a human being brush against the blade of grass before the small rodent it is waiting for, this larval mite will accept the substitute and step aboard, bringing its rickettsial passengers with it, who bring the risk of death with them. Not to the ghetto hovels though, or the pilgrim hordes, for the jungle sees only the adventurous few. It is the lonely prospector, the scout, the explorer, the soldier and

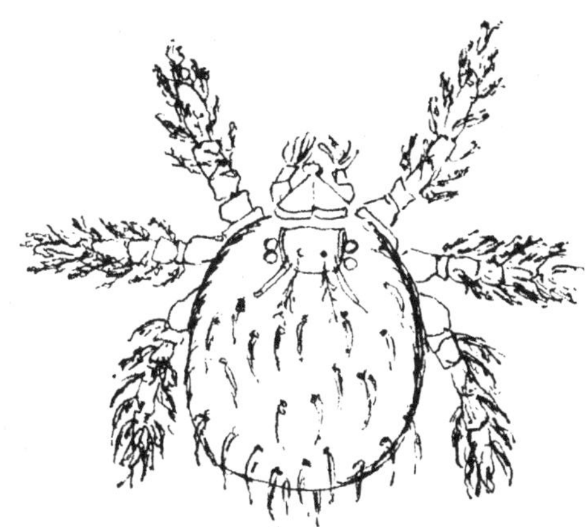

Adult mite. Trombicula akamushi.

the agricultural labourer whose rural jobs exposing them to jungle mites may succumb to scrub typhus.

The blood-sucking mite injects them as it bites, and these 0.5 micron invaders seek shelter in the cells of the inner coats of human blood vessels, preferring to inhabit the vital material of the nucleus of each cell. The cells resist, become swollen and disintegrated, and the internal strife becomes apparent in headaches, fever, chills and rashes, finally in delirium and often in death. It is a tragedy for the victim that he has intruded on the private life of a harmless mite, for mankind has no part to play in this strange cycle — the cycle from mite to rodent, rodent back to another mite and mite back to another rodent, with nothing hurtful or intolerant in the whole relationship. By contrast, the rickettsia of epidemic louse typhus has only two hosts, man and his lice, and it kills them both in millions.

Should a patient recover from any form of typhus — and there are many — immunity to that particular rickettsia is lasting and this now can be imitated artificially by vaccination.

The antibiotics are also effective in curing rickettsial diseases, which is fortunate, for the rickettsia will always be with us.

THE MINERS MITE

The engine-house is dark and still,
The life that raged within has fled;
Like open graves the boilers chill
That once with glowing fires were red;

Ah! Woeful mine, where wives have wept,
And mothers prayed in anxious pain,
And long distracting vigil kept,
You yawn for victims now in vain!

The worked out mine
Edward Dyson

Though life on a frontier tin mine at the beginning of the twentieth century was no path of roses, it included none of the inhuman drudgery which had poisoned the relationship of master and man throughout the century which had just ended in Europe. Trevor Evans knew its history well and was quite willing to exchange the life his fathers had led mining in Wales, for probable poverty in a hard land where life did at least hold a personal challenge. He crossed the world to New South Wales with his young wife.

It was just 130 years since an enterprising carpenter of Blackburn, named Hargreaves, had substituted eight spindles for the ancient single one in the humble spinning wheel to produce his "spinning jenny".

Though the hand-operated single spinning wheel and primitive loom had been accepted by mankind for thousands of years without change, once this innovation occurred it was followed quickly by many others. Within a year of the invention of the revolutionary "spinning jenny", another inventor, Arkwright, had produced something which was as fundamental to industrial progress as the invention of the wheel. To his improved spinning machine known as the "water frame" he applied a motive power other than the human hand or foot — the force of flowing water — and the mountain streams of Lancashire and Yorkshire were harnessed to give inexhaustible power to the machines. The cottage industry which had been developed under the old domestic system by the driving feet and hands of the housewives and their daughters, was doomed when the first wooden water-wheel began to turn in 1771, though it persisted side by side with the factory for another fifty years. The impact of this new power on the textile industry can be gauged by the British imports of cotton. Just before 1771 it was one million pounds per year. In 1785, by which time another driving force was beginning to be used as well, it was fifty-six million pounds, and by 1830 it was 300 million pounds.

Two years before this change from hand power to water power, three other events had occurred which were to change the peaceful easy life of 18th century England. Napoleon and Wellington were both born in 1769 and in that year James Watt took out a Patent for a steam engine. These great military leaders would have despised the humble inventor, but the hissing steam of his engines driving the noisy machinery of Europe was destined soon to make the roaring cannon of the

Napoleonic wars inaudible and forgotten. Watt's engine was seen at first only as a new machine for pumping water out of coal mines. It would allow the water levels to be kept down while shafts were dug to far greater depths. It could be used for hauling coal to the surface too, but the old system of using women and children to raise coal by climbing up ladders with baskets of coal strapped to their backs was so cheap that it persisted well into the 19th century, long after steam had been used for other mining operations.

In 1785 — steam was first used in a textile factory in place of water power. Now indeed the Industrial Revolution was getting under way. The old rural mills sited along streams were to give over to factories in towns and these towns must be near coal fields, for the cheapness and convenience of coal to feed the engines was soon apparent.

When Waterloo brought the Napoleonic Wars to an end the stage was set for the enormous industrial development which followed the peace. Throughout all these changes there was a continual increase in the demand for coal. Now it was to become an avalanche on every field in England and Wales.

Trevor's family had been miners throughout the last hundred years and had seen the death of the cottage industries and the growth of the large industrial cities with their crowded unhealthy life. It had known hunger and poverty. It had filled graves with the tuberculous, the starved, the flea-plague infected, the louse-ridden victims of typhus, the twisted and arthritic products of the wet underground tunnels, the victims of the rock-falls, the creeping poisons and the explosive gases. Trevor could imagine no dangers in a new country to equal these.

But though he may escape the hazards of the crowded city, the damp mine, and the white death which wasted and slaughtered his tuberculous forefathers, others of man's microscopic enemies flourish here as they do in the Old World from which he fled, and some of these, mistaking his inhospitable body cells for those which have given them friendly nuture for millions of years, may explode in reproductive violence in his tissues and kill him. How can the hungry mite know that the warm body brushing against the blades of grass whereon it waits for its ancient host, is a man and not a mouse?

From New South Wales, Trevor took his wife Ruth and the first of their children, little Tessa, north to Queensland, to the infant tin mining town of Herberton, where tin had been discovered in 1879, then a small, primitive settlement westward from the port of Cairns. Here he heard of good tin deposits in the little-known country further west still, and he set out to prospect there.

The parting from his family, the flinty journeys along the untrodden gullies, the lonely night camps with his hobbled mules, his ready rifle, and his night fires against the mountain cold, his laborious prospecting, were all forgotten on his return with new tin and wolfram finds located on his map, and in the good and happy days when together with a mate, the mines were developed. Donkeys, each loaded with three one hundred weight bags of ore, carried it in teams to the new village of Irvinebank. A tramway of wooden rails was opened to take it on to the batteries at Stannary Hills. The price of tin rose to $80 per ton and wolfram to twice that sum. They used only the rich ore, sometimes it was twenty or thirty percent, and if it were low, they made great heaps of it at the mouths of the shafts.

Rhys and Llewelyn were born and Tessa grew to be her father's joyous companion, prospecting with him. He taught her how the erosion of the ages had worn away the buckled hills to expose the veins of ore; how the levelling rains had carried the debris down; the unwilling tin, too heavy for any but the strongest currents spreading out fan-shaped down the mountain; how to "pan" the metal from the lighter soil and follow the "show" to the lode at its apex.

Suddenly one evil day, Tessa ran a high fever, with a fine pink rash all over her body, and wild delirium till she died a week later. They could stay there no longer after that, and besides, the price of tin was falling and the miners were leaving. The tramway to Stannary Hills was pulled up, and Irvinebank became a 'ghost' town. Only the ornamental trees remained, and still remain, in the deserted gardens; frangipani and poinciana, bauhinia and mango, memorials to the hands that planted them, flowering and fruiting in that stony wilderness, and, by their subtle alchemy, transmuting the hard rocky inhospitable soil of that decaying little mining town into luscious sugary fruits or into beauteous flowers of cream and gold, pink and scarlet, delicately perfumed.

Half a century had brought change to the big cities and to the tiny hamlets of the world. The water power which had changed to steam, had changed again to electricity and the internal combustion engine. The long paths across the world were being shortened through the skies; man dreamt of metal ships which might take him into space; the horse was discarded for the motor-car; the simple plough for the tractor.

The demand for certain metals in this new technology — ("fine word — technology!") — had jumped to unseen levels and taken their prices with them. Tin brought $1,400 per ton and wolfram $5,000 and the miners were returning to the abandoned fields. Rhys and Llewelyn came back not as their father had come, driving his mule team, but in the motor truck which on three days a week took the bags of bread and meat, the cases of tea, the crates of beer, the bundles of picks and shovels, the car tyres, machinery parts, and the school children, along the dusty road to Irvinebank. On the other two days in the week the children rode their ponies to school — learning the lessons in the sweet silence of the bush of its honest values and its unforgiving laws.

The brothers could now successfully process ores carrying as low as one half of one percent of mineral and they worked on the dumps left by the miners of forty or fifty years ago, running loads down the steep slopes on wires in counter-balanced trolleys which at the bottom hit a trip to release a peg in their floors and spill the contents into a chute which ran it down into waiting motor transports.

Again they prospered as their father had done and again illness struck as it had struck Tessa forty years before. Slowly it came on Rhys with headache, weakness and mental dullness succeeded by high fever, a rosy spotted rash and then delirium. With sudden fear, Llewelyn remembered Tessa and was thankful that he and Rhys had left their own children behind in Herberton. In his raving delirium, Rhys saw only Tessa — lovely, brown-eyed Tess, dashing out the door and along the gully to meet Father coming home from the day's work. Then she lies here flushed, fevered, restless, with Mother sponging her hot skin while Father with pale anxious face above his black beard leans over her bed. In a moment of mental clarity, he recognises Llewelyn coming with some helpers to take him to the hospital. Delirious again as he is carried down the narrow track, he hears Tess's racing feet and his own in the rough games of their childhood, with every dislodged stone that scampers down the steep hillside.

Tess, gone these forty years! Why should she come back so vividly to his congested brain? And what curse lies on these hot hills that seizes the young girl and the grown man in its pyrexial power and strikes them both down in sweating delirium?

A very small enemy indeed, about half a micron long and just within the limits of microscopic vision, though it has a long name — Rickettsia tsutsugamushi. Once thought to be transitional forms between viruses and bacteria because they partly share the characters of both, the Rickettsia are now recognised as true bacteria which possess such rudimentary enzyme systems that they must live as total parasites within the cells of their hosts which usually are lice, ticks, fleas and mites. Human body lice when infected by them have carried the sentence of death by typhus fever to millions of people in cities, armies, jails and ships, whenever there is crowding. Ticks, fleas and mites have killed their lesser millions in the less populous places of the Earth, unfortunately mistaking man for the other mammals or rodents from which they usually obtained their meal of blood.

In their prospecting sorties along creeks and gullies, Tess and Rhys had each brushed against a waiting larval mite which had climbed onto their bodies and bitten them. So small that a dozen of them can fit on the head of a pin, neither their presence nor their bite would be noticed, but these larval mites, the accomplices, sheltered and

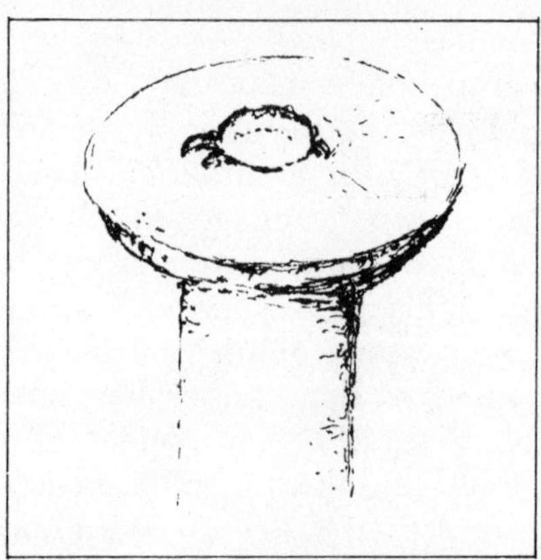

Larval mite on head of pin.

sustained the true murderer, the fatal rickettsia of Scrub Typhus. The mites themselves inherited it through the eggs from which they arose, even to the second and third generation. When the assassins can hide in multitudes, invisibly, in the invisible egg of a barely visible mite, appearing also in the new mite arising from this egg and again in the new egg arising from this new mite, it must surely need the detective genius of an expert to apprehend it.

Where the mite bit, it left rickettsia to form an ulcer. Then, grown to incredible numbers, these entered the cells lining the blood capillaries in heart and brain, and disorganised the action of both so that headaches progressed to delirium, and the regular heart rhythm fell to a slow unsteady beat.

These changes which had overwhelmed Tessa forty years before, would have killed Rhys too, for in forty years, perhaps in forty million years, the parasite had not changed. In incalculable numbers through countless transfers, the mite, the rickettsia and their rodent hosts had established mutual tolerance, but Man, the newcomer, not knowing the secrets of peaceful co-existence could only kill his attackers. Where his quarry was too small for bow and arrow, grapeshot or grenade, he could use chemicals. Rhys was given chloramphenicol.

His unseen assailant was destroyed and the defensive cells in Rhys' body had learned a lesson they would never forget. Whenever this particular rickettsia might attack him again, these cells would retaliate so quickly that the invader would be destroyed at once. He was henceforth immune. Perhaps in the far future, men, like the small rodents of the scrub, will be immune too, or perhaps together with the lice which live on him in other parts of the world, he will still die from rickettsial invasion, for changes in the characteristics of those body cells which defend us, evolve, like other biological changes, only with the ages.

12

GONORRHOEA

The little kidney shaped unwelcome visitors to the human genital tract — Neisseria Gonorrhoea — only 0.8 microns in diameter live together in pairs sometimes even inside the defensive cells of the body whose function should be to destroy and not to harbour them. In spite of all the devious ways erring Man claims to have acquired them, they are so easily killed by adverse influences such as drying that they are transmitted only by sexual contact, or unhappily to the eyes of a new-born babe from an infected mother.

If the Gods could annihilate both Time and Space, it is said it would bring happiness to many distant lovers. And if some other gods with a more personal responsibility towards lovers could annihilate the chances of unwanted pregnancy, the sufferings due to syphilis, the sordid symptoms of gonorrhoea, the risks of four or five venereal virus infections, the filthy body louse, and the itchy mite of scabies, the prostitutes might be more prosperous, and the promiscuous male might have less to worry about.

But the Pill, Penicillin, and Permissiveness have not co-operated to bring about the Utopia which was expected of them.

The moralists may moralise and the preachers preach, but a business with its beginnings so far back in Time, and with so certain a ''consumer demand'' is unlikely to fail because of some occupational risks. There will always be gonorrhoea until some dramatic cure is found. At one time penicillin seemed to supply this, but it is not now so effective because the gonococcus like other organisms has shown that it can develop resistance and outlive antibiotic therapy. It is now causing world concern about its future effect on public health.

Something between fifty million and one hundred million cases is the estimated present global incidence of gonorrhoea. The statistician has

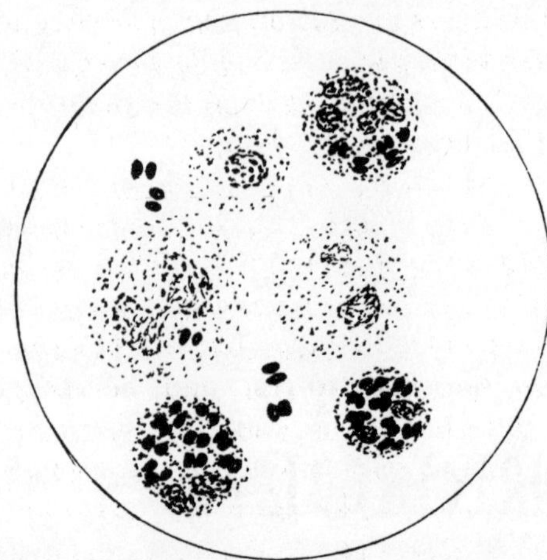

Neisseria gonorrhoeae being ingested by white blood cells.

an impossible task of course, for the person who contracts the disease, often neglects treatment.

Being totally parasitic, the organism is obliged to live right within other cells to obtain the nourishment it cannot prepare itself. It therefore burrows into the cells of the soft mucous surfaces causing inflammation and, spreading back along the adjacent structures from the venereal site where it first began, it carries inflammation to the bladder and its urinary apparatus and to the reproductive organs, often resulting in urinary obstruction and sterility. It may spread to other parts. In the eye it may cause blindness, in the joints, arthritis, in the brain, meningitis, on the skin, dermatitis. It is a disease which is usually easy to diagnose and treatment, if early and thorough, is quite effective.

LOVE ME, SAILOR

Wi' lightsome heart I pu'd a rose
Upon its thorny tree;
But my fause lover staw my rose,
And left the thorn wi' me.

Ye banks and braes.
Robert Burns

Noon lies heavily on the city of Naples. The shops and offices are shut. The hotels and houses have closed the green lattice shutters on their windows and gone quietly to sleep. Traffic has almost ceased. It is siesta. A few children still play in the streets, and a few policemen watch from the shade for the lawless, if lawless there be in such mid-day heat. There are a few resting buses in the magnificent square, Piazza del Municipio, that faces the harbour.

At the Statione Maritima where the liners berth and the Customs Officers challenge every vehicle entering under the wide arch from Via Marinella, the tourists sit drinking wine or tea in the dainty bars with a view of the shimmering bay. Occasionally a street car runs along the harbour front and away to distant suburbs. The drivers of the droshkys put their horses in the shade to rest, and, bending down, turn on the water taps to fill the stone basins sunk just below street level, which the city provides for equine refreshment instead of ugly troughs.

Water spurts lazily in the fountains in those long narrow rock pools which surround this large Piazza, separating its grassy lawns from the hot asphalt of the streets.

There are a few tired-looking bees about the dull drooping flowers whose very perfume seems to be awaiting the cool freshness of the evening air.

The shoe polisher and the post-card vendor have picked up their little wooden frames and taken them home till a cooler hour. Even the furtive seller of contraband watches and fake jewellery has found business elsewhere for he would be too obvious in the deserted streets.

On the hills to which this lovely city rises there is a soft heat-haze and south across the blue Bay, the double hump of Vesuvius makes an ancient and beautiful picture. A tug-boat loiters nearby for there are some ships that sail even at mid-day, and their final needs may bring a few trucks slowly over the hot surface of the terminal.

Except for these, it is silent, still, and restful, and even the pigeons know it is siesta.

But as the sun sinks westward to cool itself in the Mediterranean, life runs back into the Neapolitan scene. The pigeons start to coo and shake their feathers. The sleepy longshore man raises himself stiffly from the deep barrow he has used as an armchair for his slumbers in the shadow of the elevated gangway. The droshky driver rolls back into the shade which has deserted him as he slept on the lawns of the Square, sits up and speaks to his horse, almost asleep too under its sun-hat with its nose touching the ground. The weary animal starts at the sound of his master's voice, tosses its head, stamps its feet a few times, yawns, and is ready to drag its glossy vehicle over the cobbled streets again at 5,000 lira a kilometre for tourists and 1,100 for the locals. The haze disperses. Tug boats hoot across the harbour, buses and street cars follow each other in increasing numbers, the green shutters are thrown back, the shops re-open and the crowds return.

Fountains gush up merrily all over the Piazza and little naked children jump into the pools around them to sport till nightfall. The boot-black sets up business again at his corner, and while he polishes your shoes, the watch-seller takes advantage of your immobility, "I give you da great bargain, Sir? Yes? No? Then how much you like to pay?"

The shoe man grins at the persistence of the vendor who you both know is trading suspect goods, and in your anxiety to get away charges you for each shoe what he should charge you for both. The vendors have returned with the picture-cards, the jewellery, the gaudy scarves. To what remote corners of the Earth have they not carried Vesuvius and the Bay of Naples!

Then the bars open and wherever tables can be set up on shady sidewalks, drinks are served. The policemen, having found no evil-doers, patronise the bars, paying for their drinks if there is no other way for an obliging policeman to reward the proprietor.

"Naples" wrote Shelley, "which ever pantest naked beneath the lidless eyes of Heaven."

Today, Heavens lidless eyes have warmed the waters of your Bay and set the juices ripening in the vineyards of your hills.

Come, sip their wines, talk the evening away, dance the Tarantella, sing the songs of Naples . . . O Sole Mio . . . Come back to Sorrento . . . Funiculi Funicula.

The sailors come ashore from the visiting ships to share the gaiety of this happiest of all Italian cities. And, of course, Lucia and her kind, come to share the sailors.

Lucia is big, blonde, brazen, with a fine figure. Her dress now is of her vocation — hatless, with a black Hong-Kong wig, a tight red silk blouse, cut low, the miniest of mini skirts, black panty-hose, black high-heeled shoes, cosmetics laid too generously on a once pretty face.

She is alert for custom, and walks briskly around the Piazza from the Castel Moro, that medieval fortress whose massive slotted walls and deep dry moat tell an ugly story of feudal times; up to the Galleria Umberto, across the Piazza and down the hill again to the Statione Maritima. The sailors ignore her greeting and sip their wine, but Lucia is quite prepared to wait. She has seen it all before and knows what a seductive Neapolitan evening and a little stimulating alcohol can do. So, on the prowl and round the beat again. There may even be someone lonelier and wealthier than a sailor wandering by himself in the Piazza del Municipio this lovely evening.

Joan Murray sat in the empty nursery looking with an intense longing at the treasures she and Harry kept stored there, the clothes she had sewn with such pleasant anticipation; each completed garment adorning one of those dream babies she and Harry had dressed so often that they had taken on a kind of reality.

The toys Harry had made were taken just as often from his hand by the same eager baby fingers as fast as he could make them. The cot, chosen with such care for the new owners-to-be who would one day burst into their lives from nowhere, standing in starchy newness in its window corner, already held in fancy the little family they planned to raise.

But strangely no little family came to be raised, and the dresses lay in their boxes among the lavender sachets, the toys sat forlornly on the shelves hiding their delights in the slowly gathering dust, and Joan's visits to the nursery were sometimes accompanied by a few desperate tears.

When doubts could no longer be hopefully explained they decided that Joan should see Dr. Melville who had cared for her since childhood. He found her a perfectly healthy young woman. Then why, she asked him, was she childless after years of marriage.

''My dear Joan'', the doctor told her, ''childlessness is not due only to female defects. You have a husband and I want him to see me also.''

Like Joan, Harry was normal to physical tests, but, as a new life even in the largest animals, commences with a microscopic cell, only a microscopic examination will show if these cells are present, absent, or defective, when they have not revealed themselves by reproduction. It showed that in the semen from which Harry had hoped for sons and daughters, there were none of the seeds of life. The three hundred million or so spermatozoa which should have been there were entirely absent. Nothing Harry could do would ever fill the vacant cot in that empty nursery or furnish the baby fingers to play with the toys he had so fondly prepared.

Distressed, Harry asked why he, a healthy young man always A1 in the Navy and never really ill in his life should be sterile when other fellows with half his virility reared big families.

''If you don't mind answering some personal questions Harry, we may find out. You were in the Navy?''

''Yes, doctor. And honourably discharged with a clean record five years ago.''

''You visited foreign ports?''

''Yes, of course; that's what the Navy's for.''

''And there were girls in those foreign ports. Did you visit them too?''

To the doctor's sympathetic questioning, Harry told the story of that evening in Naples, the wine, and the persistent Lucia, but though admitting that the sailors had at last accepted her company, he claimed none had suffered anything except perhaps the mildest inconvenience — a little urinary scalding, a slight discharge, a glued-up eye, all common disabilities in tropical regions. He had always thought that venereal disease would be accompanied by something distressingly severe. The doctor told him that where there is promiscuity, there is always the risk of infection, and that gonorrhoea can range from a quickly fatal disease to a mildly troublesome infection such as he had described. The germs which cause it are only about eight ten-thousandths of a millimetre in diameter, which Harry would realise from his knowledge of naval gunnery, was an invisible measurement. They can penetrate the living cells on moist inner surfaces and in his case they had reached the sperm ducts to inflame and obliterate them. Dr. Melville explained the extreme intricacy and therefore the extreme susceptibility of the reproductive cells — the germinal epithelium budding off millions of mobile spermatozoa each containing chromosomes which themselves contained millions of genes, and all the other complicated mechanisms of inheritance till Harry realised that such complex machinery could very easily be damaged.

Joan and he, the Doctor said, might have been the victims of one of life's worst tragedies, as babies born to parents unaware of their infection may sometimes be born blind. There were in America at present, twelve thousand such children.

"And I'd have made the poor kid blind." Harry was shocked at the possibility.

"Not you Harry. The wine you had at Naples. It made you blind, didn't it? One other thing. If you haven't told Joan all about Naples I just want you to know that a bad attack of mumps can also produce sterility. There can be no good purpose in letting the past spoil the happiness of the future.

Though you and Joan will never have a family, there are children of other parents who can be adopted and loved just as your own."

Thanking the Doctor, Harry went homewards thinking sadly how the lives of men and women can be so deeply influenced by living organisms of a size and volume so small as to be invisible and unknown to all but the few who study them.

So the little child that was wanted and never came, gave up its empty cot in the silent nursery to the little child that came and was not wanted, and often as Harry watched Joan's loving care of it, he wondered how many other nurseries around the world stood empty because ships sailed into that beautiful blue bay and grapes ripened on those sunny hills.

13

CHOLERA

The Cholera vibrio is really an Eastern gentleman and he is no more welcome there than in the West to which he makes frequent journeys, for he is a ready traveller and has visited most countries, especially the warmer ones. He has found his way through the caravan, the pilgrim, and the trade routes of the world, and along the major shipping lanes even to America and Australia. He is now also an experienced jet-setter, dropping in unexpectedly on air-line re-victualling bases. He has always loved the Army and has joined in most campaigns, and he is quick to appear wherever there is a catastrophe. But his ancient home has been India and South-East Asia. He was definitely identified and labelled by a German bacteriologist, Koch, in 1883. He belongs to the Vibrio family, small active bacilli, and he is called Vibrio comma, because of his curved shape. He is one and a half microns long and one third of a micron in diameter, but though so tiny, has killed millions and will kill millions more. He contaminates many foods but the main epidemics are caused when he gets into a public water supply. When swallowed he sets up a toxic irritation of the bowel wall which disintegrates and causes diarrhoea so copious that the patient becomes dehydrated and the whole circulatory system may collapse from loss of body fluid. About one half of all cholera patients will die if there is no replacement of fluids and minerals. With correct treatment, all but a few now recover.

Why, of all the animals available to it, should this organism choose man and man alone, in which to live its loathsome life? A few very close relatives do infect cattle and horses, the vibriosis they produce being mostly of the genital tract, resulting in abortion and infertility, but the vibrio of cholera has never been known to have any host but the human host and it is purely a local disease of the intestines.

What conditions do they find in man that they can find nowhere else? Where too do they hide to wait from epidemic to epidemic for chronic human carriers are said to be unknown? That is another mystery for the microbiologist to solve.

"OF OLD, UNHAPPY, FAR-OFF THINGS, AND BATTLES LONG AGO"

Big fleas have little fleas,
Upon their backs to bite 'em
And little fleas have lesser fleas
And so — ad infinitum.

Dean Swift

Captain John Wickham took a long look back at the old family home in Surrey, its grounds so lovely in the Spring of 1854. The trees had never seemed more beautiful. Oak, elm, and sycamore bursting into leaf; peach, apple and plum in full blossom or with young swelling fruit; rose, daffodil and rhododendron adding fragrance and colour.

England was so beautiful in Spring — dash these foreign wars, Sir!

The old coachman drove slowly towards the railway station as he had done so often when John Wickham was going back to school after his midsummer holidays, though today both were unusually silent.

"Goodbye Sir", he said, bareheaded, as he left his young master. "And we hope it be soon as we drive you home again, the horse and me. When you've beat the Russians!"

Grenadier Thomas Slater kissed his wife and children, unclasped the youngest from around his neck, and passed her sobbing to her grieving mother, shouldered his pack and his weapons, gazed fondly at the row of houses in the dull street in London where he had made his home and marched sadly around the corner with many fondly waved farewells.

The emptiness in the ancestral home in Surrey and the terraced house in London were matched that night in many sad homes in England, for the Army was leaving for the Crimea. Brave and soldierly they looked next day as, in full battle order, they marched out of the courtyard of Buckingham Palace, down the Mall and through Trafalgar Square on their way to the troop-ships. There were no better troops in Europe! Well drilled and disciplined, inspired by the memory of Waterloo and sure of their ability to thrash the Russians and return to their families before the summer ended. They had the might and prestige of a powerful Empire, respected around the world, to support them. The bands played and the crowds cheered their magnificent fighting men!

By the time the transports reached the Mediterranean, the early confusion had given way to Army order and discipline. The horses were tethered between decks in two long lines facing inwards, their head-ropes tied to a thick cable running the length of the deck at chest height, and their heel-ropes to a similar cable lying on the deck behind them. The waggons were lashed on the decks and filled high with the impedimenta of Cavalry and Artillery units. The guns were safely stowed, but the troops found what comfort they could lying between the waggons on deck or down in the holds, rolled in their blankets with their boots for pillows.

They were to land at Varna in Bulgaria to relieve the besieged Turkish garrison but they had scarcely disembarked when cholera appeared and by the time they had been taken on board again, one thousand cases had to be sent to the base at Scutari where old Turkish barracks were to be used as a hospital. There were not then enough transports to take the entire Army across the Black Sea from Varna to the Crimea. Thirty thousand men were embarked but most of the equipment was left behind, so that when the battle of Alma was fought a week later, there were no medical stores to treat the five hundred battle casualties. With a thousand new cholera cases these were to go back to Scutari, where medical supplies were as scarce as in the Crimea.

"The Director-General of Medical Services in the Crimea is informed by the Commander-in-Chief that Her Majesty's Transport 'Kangaroo' is at his disposal for the transfer of hospital cases to base at Scutari. He will place battle casualties from the recent battle of Alma first in the transport, followed by such medically ill personnel as he thinks fit. H.M. Transport 'Kangaroo' is fitted out to accommodate 250 patients."

No longer erect and soldierly, their uniforms dirty, blood-soaked and adherent, smelling of pus and putrefaction, perhaps pursued by hosts of flies, the wounded are always pathetic. They are most pathetic when, besides being wounded, they are neglected and forgotten. They were rejoicing and ringing the bells in England in honour of their glorious troops. The politicians and leaders at the War Office were congratulating each other, for the Crimean contracts had been expensive and political prestige depended on a quick and successful campaign.

England had reason to be proud of her soldiers; these magnificent

men who had won the battle of Alma in a blizzard. Hang out the
bunting! Ring the bells! No enemy can defeat them.

No enemy?

None perhaps but cholera.

These little vibrios had been in most campaigns. They were seasoned
old veterans from the wars of vanished Empires, but the leaders at the
War Office did not give them a second thought.

The glorious troops then met with unexpected opposition and
suffered severe casualties in battles fought in fog and cold. The
wounded straggled down the treacherous slopes to the sea in bitter
weather — unaided, and without supplies — without morphia,
dressings, bandages, splints — often without food or water.

Grenadier Slater discarded his pack because he had been ordered to
discard it on those grim hills and with Edward Tomkins, Lance-
Corporal, also severely wounded, finally completed the slow painful
journey to the tent hospital and was allotted to the ''Kangaroo''.

Her Brittanic Majesty's Transport ''Kangaroo'' fitted out to
accommodate 250 patients, was loaded with one thousand five
hundred, of which five hundred were wounded, and one thousand were
cholera cases. By the time the slow, heavy little ship dropped anchor at
Scutari all on board had cholera.

Grenadier Slater, groaned with pain as he was dropped overboard
from the ''Kangaroo'' into a small boat and landed on that dreadful
beach, for there was no landing pier. The hospital had received no
notice of their arrival, but had little means to provide for them had it
been told. Clutching their blankets, their only possessions, for they had
been ordered to throw away their packs on those difficult Crimean hills
in defiance of Army regulations, those who could, crawled up to this
''hospital'' with no beds, no nurses, no equipment, little food or
medicine, and few doctors, but with an abundance of lice and flies.
Those too stuporose with disease or too severely wounded to walk, were
carried roughly in ''litters'' by Turkish orderlies, who, having for some
reason deserved punishment, were doing this work as a penalty and
were resentful and careless.

Now, beside their untended stinking wounds and battle dress sodden
with blood and filth, their bodies were dehydrated, their eyes glittering
with the high temperature of extreme illness, their faces grey with pain,
their fevered sweating bodies unwashed and lice-infected; and,
reaching the bare, cold hospital, they lay down on the stone floors
rolled in their blankets, and waited to die. The piteous heap that had
been Grenadier Thomas Slater huddled silently on the floor of this
great death-house with Ted, his cheerful Yorkshire friend, the Lance
Corporal, dead for some hours of cholera. Ted had asked him to write
to his wife when he reached England, but Grenadier Slater knew he

would never see England again. Like so many others he saw carried out daily, he and Ted would be buried on this bleak foreign shore.

There had been talk lately of English nurses being brought out by a Miss Nightingale, but the officials were worried, the doctors jealous, the patients indifferent, and the matrons of England scandalised.

The crowds had cheered the troops in London but had left them to die in thousands, after Alma, after Inkerman, at Balaclava, at Varna, at Scutari and at Sevastopol. There were more men ill than in the ranks. Of those sent to hospital 73% died there.

Well — it was just part of a soldier's life. If there was no water to wash the soldier's wounds, or none for the cholera patient to drink, he must remember that the tender-hearted never built an Empire. The sick and wounded must do the best they can.

It was then that Russell, a war correspondent, sent some of these deplorable facts to John Delane, Editor of The Times. The national conscience, which had been faintly stirring, came to sudden life. The Times Fund was established to provide for the health and comfort of the private soldier. Miss Nightingale and her forty nurses left for the Crimea to begin a long and difficult battle. Gradually from the incredible filth, the bungling disasters, and the woeful tragedies of this calamitous campaign, the prejudices against the employment of women in the Army and elsewhere were overcome and today's concept of hygiene and prevention in hospital and medical practice slowly evolved.

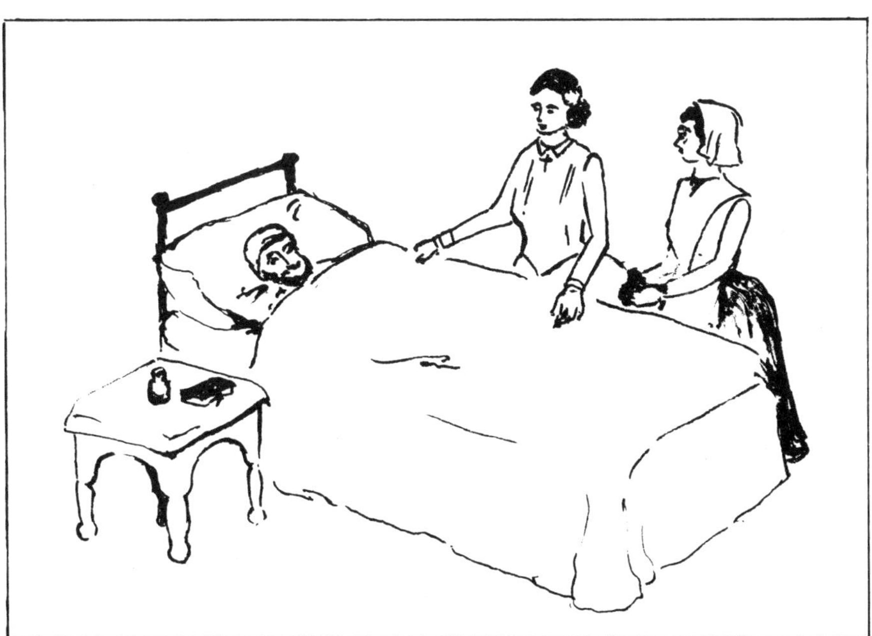

"If it had not been for Miss Nightingale", said John Wickham, "thousands of us would never have returned to England".

"What did the doctors there think was the cause of cholera, John?" his sister asked.

"No one knew, Jane. Here comes Dr. Gray though, so you may ask him. Perhaps they have lately discovered the cause."

"Cholera, Jane?" said the doctor, accepting a cup of hot tea after inspecting John's healing wounds. "If I could answer that question, the Queen would send for me. When I was a student, it was said to be caused by the climate because it is so common in hot countries, but we had an epidemic here in England twenty years ago and fifty thousand people died, and last year we had cholera here in Soho — and some of us have traced many of the cases to water from that vile pump in Broad Street. I'm sure now that foul water is the cause. There has been no cholera since we locked the Broad Street pump. It seems the disease may be caused by living things too small to be seen."

"I heard the medical officers in the Crimea discussing this Dr. Gray, but they said if it is due to these small living things, why does it kill only men, while dogs and other animals drink the same water with safety. Why does it kill the one and not the other?"

Dr. Gray, having no answer to this question, pulled out his gold watch and asked if they might be so kind as to have a servant ring for his coachman, and as he sped down the driveway, he thought deeply about Wickham's question. Why indeed? This seemed to make nonsense of all the grand theories of living germs of disease.

The nineteenth century was nearing its close, before it was known for certain that there were living invisible things more deadly than grape-shot or bayonet.

It was difficult for Dr. Gray and his colleagues to accept the theory of living invisible organisms of disease. It would have been totally incomprehensible that these invisible predators were subject to like depredations from still smaller living organisms. Yet these bacteriophages — a word which literally means devourers of the bacterial cells are now well known and are used in diagnostic bacteriological laboratories to identify some of the bacteria which they destroy. A bacteriophage which attacks the cholera vibrio seems to be a vital factor in checking epidemics of cholera. It is when the cholera organism finds its way into a person totally unprotected by antibodies in his blood or "phage" in his bowel because he has never been infected before, that it produces disease; and 30,000 English soldiers newly arrived in the East would provide it with such a virgin population.

Having gained entrance, it multiplies to uncountable millions and by its destructive toxin irritates the bowel wall, which then liberates fluids into the bowel producing a diarrhoea in order to remove the invader and its irritants. But the little vibrio multiplies so fast and its toxins have such power to increase the permeability of the blood capillaries of the bowel wall that vital fluid continues to be drained away from the circulatory system into the bowel and lost, till not enough remains to fill the blood vessels, and maintain blood pressure, and the depleted vessels collapse. Should the cholera patient be a Crimean soldier on Her Majesty's Transport ''Kangaroo'' losing blood from a wound as well, he will have to fight for his life on two fronts.

He will be pale, cold and breathless with a heart beating rapidly to pump the insufficient blood supply round, more often; he will be thirsty and calling for water for his dessicated body — and in the hospitals at Scutari and on the beaches of Crimea, there often was none. As the fluids of his blood are constantly drained away leaving the cells behind, the blood will become so concentrated that his kidneys will fail to filter it.

So by a fierce dehydration based on the power of a toxin to increase the porosity and permeability of the capillaries of the bowel wall, the patient will die in a manner much resembling death from thirst.

In a modern hospital dehydration would be treated by running into his veins a continuous supply of chloride, glucose and carbonate solution to replenish his fluids, fill his collapsing veins, control the balance of the minerals which keep the cells distended and the electric systems working, neutralise the acidity which the imbalance has generated, and provide his infected body with nourishment.

Vibrio comma is partly controlled too now by some antibiotics but the treatment of dehydration is the main need and it is unfortunately true that in many parts of the world where Asiatic cholera is severe, the facilities for such care on more than a limited scale do not exist. So until people practice simple hygenic measures for prevention, Asiatic cholera and some similar diseases will remain dangerous killers, both in the East and in the West.

14

TRYPANOSOMAIASIS

If you took a drop of blood from the nearest rat in whatever part of the world you may live and examined it under a microscope, it is very likely you would see a slender tapering organism about thirty microns long, wriggling through it. It would move by the propelling action of graceful rhythmical waves which travel along a thin membrane running the full length of the parasite, and because of these motions known as an undulating membrane. At its front end there would be a long free flagellum like a whip, which would continue on to form the outer edge of this undulating membrane. You might see it split lengthwise into two, for that is how it reproduces itself. It would wriggle and swim amongst the red cells and you would be fascinated, watching it.

You would then have seen a Trypanosome, a member of an immense group of microscopic animals living their complicated lives within one single cell and known as Protozoans. Trypanosomes live alternately in flies and bugs and in men, cattle, horses and rats, commuting between the two hosts, and assuming about four different forms in the process.

The wicked witches of the fairy stories must have known a great deal about physiology, biochemistry and molecular biology, to change frogs into princes so easily, for so far Science has not revealed to us the intricate processes by which these stately aristocratic swimmers in the body fluids can change into dumpy common little plebians with no undulating membrane to lend them grace and motion, or even into rounded dots with a diameter about one tenth of what was once their length, entirely without membranes or flagellae, immobile and crowded together in clumps within infected cells. And like Cinderella, they can change back again at the right time. They are unique in that they seem to possess a kind of second nucleus to the cell.

Though trypanosomes are present in many parts of the world and though seven species are found in man, it is only in Africa where they

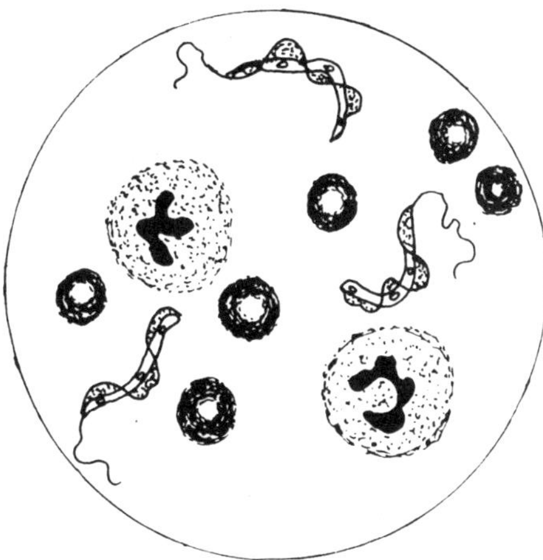

Trypanosomes in blood smear. Cause of African Sleeping Sickness.

cause African sleeping sickness and Latin America where they cause an equally fatal heart disease, that they seriously affect man or animals. This may be due to changes induced in them by their cycle in the African tsetse fly which lives only in Africa, or the South American bug which likewise is found nowhere else but in South America and which are in each case the carriers of the parasite.

It is currently estimated that seven million people are affected by this trypanosome in South America, and the numbers afflicted by ''sleeping sickness'' in Africa could never be known. It also causes deaths amongst cattle and horses there, making large tracts of Africa unsuitable for raising cattle and so further reducing an already limited human protein diet.

The trypanosomes have been carried to many places in the bodies of African slaves, but have not established themselves in other parts of the world. The slave traders were well aware of the clinical signs of the disease and refused obvious cases.

So the slave in his bondage in a new land was at least freed from his dangerous master, the trypanosome, in the old one. The absence of the fly or the bug will explain the absence of the diseases, but what can explain the curious fact that the trypanosomes live harmlessly within the rat or the antelope, the fly or the bug, but rapidly destroy man and his horses?

SITATUNGA

''And now we turn to Nature. All these years we have lived beside her and have never seen her; now we open our eyes and look at her. The rocks have been to us a blur of brown; we bend over them and the disorganised masses dissolve into a many-coloured, many-shaped, carefully arranged form of existence . . . The flat plain has been to us a

*reach of monotonous red. We look at it and every handful of sand starts into life . . .
The bitto flower has been to us a mere blur of yellow; we find its heart composed of a
hundred perfect flowers, the homes of the tiny black people with red stripes, who move
in and out of that yellow city. Every blue bell has its inhabitant. Every day the karoo
shows us a new wonder sleeping in its teeming bosom."*

The Story of An African Farm
Olive Schreiner

Only a few slender honey-brown biting flies not much bigger than
their relatives, the common house-fly, flitted about in the dappled
sunshine and shade on the bank of a stream in the mid-day heat of the
African forest. Aristocratic home-loving flies. They did not spread
around the world like their cosmopolitan cousins but kept to a restricted
equatorial band of their native Africa about 15° on each side of the
Equator. Despising the filthy feeding habits of the house-fly, they lived
on warm vertebrate blood which they drew with surgical skill, cutting
through the skin with the fine stylets of their proboscis and the rasps on
their pointed lips. Then they settled on the vegetation with their wings
neatly folded and not held vulgarly at an untidy angle as other resting
flies did. In their sheltered sylvan life, they lived to a greater age than
most flies moving greater distances about their domain and drawing
their blood meals from many animals including Man.

Close by, hidden by the shrubbery he had arranged around him,
Ngami lay motionless and almost invisible in the changing pattern of
light and shade, cooled by the light noon-day breeze which blew his
scent away across the river to smother it in the forest beyond. Some of
these flies alighting on his naked skin took a few cubic millimetres of his
blood, but he never flinched. The hunter must fade into his background and remain rigid if he is to surprise his timid quarry. Between
the hunter and the hunted, the eater and the eaten, there is constant
war, and constant wariness.

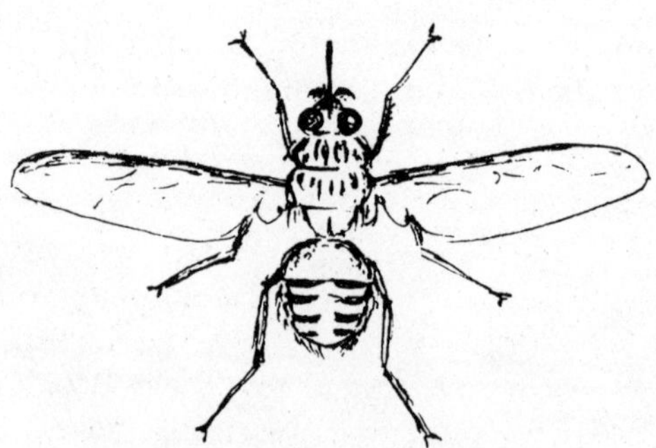

Tsetse fly. Carrier of the trypanosomes which cause African Sleeping Sickness.

The antelope herd approaching with cautious sniffing noses, ears flicking and eyes turning towards every point of danger knew nothing of their concealed enemy until the arrow struck, and a young doe, bounding high with pain and fear fell with convulsive dying kicks down the bank towards the water she had come to drink.

Poor gentle little antelope, so harmless, so perfect, so beautiful — they call you sitatunga — a victim to the thirst that brought you cautiously to the stream and the silent arrows, to be slaughtered so that Man might have food. Perchance though it is now Man's turn to die, for as you go, little antelope, to the village feast, revenge is going with you. Those honey-brown flies will yet redress this grievious injustice, for when you last came here to the stream these tsetse flies, ravenous, like men, for blood, drank yours, as today they drank Ngami's. With yours went the elegant contaminating trypanosomes, 30 microns long, swarming in the fluids of your graceful body. Over the millenia of your common evolution you have learned to live together so that they did you no harm, though reproducing in millions in your blood.

But when some of your trypanosomes were drawn from your warm blood into the cold gut of the bloodless fly, these adaptable parasites merely assumed a less elegant form, and proceeded on an orderly circuit of their new host, until in twenty days, they reached the salivary glands, distending them with their increasing numbers. Here, their journey halted, and they lay poised, waiting for the fly to feed again and allow them to escape with the lubricating saliva which assists the penetration of the slylets, back to the home of their ancestors, the warm animal blood stream.

What force drives these comely parasites, imbibed accidentally by a hungry fly to set off to a regular pattern and time-table, through seven anatomical structures of their new hosts to reach a strategic position in the salivary glands — the only possible one — from which they will get the chance, figuratively to return to the land of their birth? What physical laws direct their course? Could this be a remnant of their early history? Were they first intestinal parasites of the fly, accidentally infecting the animal's blood, or parasites of blood infecting the insect's gut? As only one per cent of tsetse flies can be infected by feeding on the blood of animals infected with trypanosomes even under the most suitable conditions, it seems that they are indeed parasites of the insect's gut and are only beginning the evolutionary experiment of inhabiting the blood stream of certain animals.

With all its skill however, this little parasite may, like mice and men, have chance encounters with Fate. Today, its host the tsetse fly, chose to bite a man and not an antelope. Now, injected into human blood, which, unlike the gut of the fly, should resemble its former home in the blood of the antelope, it finds that something is lacking; and though it lives harmlessly in the antelope, and survives the transfer to the body of a fly without damage to either, yet in man it proclaims total war . . . war to the death, for they will all perish together.

We say that man has no immunity to trypanosomes which is a way of admitting that he is still a few million years too immature to know how he and other beings can live together without mutual slaughter; that he is a newcomer of whose presence on Earth there is no evidence for some four hundred million years after the lowly cockroach left its first records, and the trypanosomes would be old inhabitants in geological time before the cockroach existed.

It took one year for Ngami to die.

In one week, the trypanosomes had broken from the bridgehead at the bite, where an automatic increase in the flow of blood had flung a ring of defensive cells around them. Then, swimming freely in the circulating blood, for, like all protozoans they are really aquatic creatures, they explored every tissue, apparently damaging nothing.

But the aggressive defensive cells in Ngami's body recognising the proteins contained in the trypanosomes as foreign, and therefore enemies, challenged them and reacted with them, so that a raging war began, protein against protein.

Probably because the trypanosome needed to vary its physiology to make it acceptable again to the tsetse fly after its aberrant sojourn in human blood, it continuously changed the chemistry of its proteins, and the defending cells, responsive to these slight changes, produced successive crops of changing antibodies to destroy them. These antibodies built up to an extraordinarily high level, and defeated their own purpose by themselves irritating the healthy tissues.

The violence of this inward war showed itself in headaches, fever, rashes, swellings and pain. The swollen lymph glands stood out, at first soft and tender, and later hard and rubbery, especially along the back of his neck. Had Ngami lived in slave trade days, these would have reduced his price or made him unsaleable for it was the hallmark of death. The concept of ''planned obsolescence'' which burdens commerce today had no place in the slave-traders philosophy.

Then the parasites reached the fluid which surrounds, protects and nourishes both brain and spinal cord and travelled along their tiniest vessels.

The inevitable result of this unnatural irritation of such specialised cells was their degeneration and their replacement with the coarser tissues of chronic inflammation. The brain cells slowly ceased to function.

Ngami, that great hunter, became more and more languid, apathetic, lethargic. The quick steps of the agile young man who stalked the antelope along the forest streams became the weary shuffling gait of senility. The alert brain was overcome by sleepiness from which arousal became increasingly difficult. He found it hard to speak, harder still to take food; till at last his emaciated body fell into the coma of starvation and brain destruction from which there was no awakening.

The Trypanosomes had won a Pyrrhic victory. They died with their host!

15

BACILLARY DYSENTERY

The alimentary tract of man deals with many kinds of food from many diverse sources. No wonder then that unwanted material in the form of microscopic life is sometimes swallowed too. Not that the alimentary tract is unfamiliar with microscopic life in fact it swarms with it, millions to the gram. Some of its many inhabitants are part of the nutritional system manufacturing vitamins and helping in the digestive processers; some are so universally present that they are used as an index of faecal contamination of public water supplies. But some species are no friends to the alimentary system.

There are for example the dangerous mobile Salmonellae including in their family over 1000 strains all looking exactly alike and differing only in their antigenic patterns. The Salmonellae have been found in many types of food; in milk and all its products, eggs and meat, and in such unlikely vehicles as dessicated coconut and food colourings.

One species of Salmonella, the bacillus of Typhoid fever starts its fatal attacks from the bowel and then invades the blood.

The Cholera vibrio, patron saint of pilgrimages and so well known in historys harrowing epidemics, attacks only the human bowel but survives in unsuspected streams. The Shigellae, minute organisms 2 microns by 0.5 micron, when swallowed in an infecting dose of at least 10,000,000 attach themselves to the lining of the intestine, multiply, and induce severe inflammation, high fever with pain and diarrhoea, ulceration and perhaps perforation of the bowel in the same way as Cholera, not therefore making much contact with those cells which involve the production of antibodies. For this reason vaccines are of little value against dysentery and Cholera but effective in the prevention of Typhoid fever. Bacillary Dysentery is highly contagious and has rivalled Cholera and Typhus in fatal epidemics whenever there is a concentration of people in less then ideal conditions.

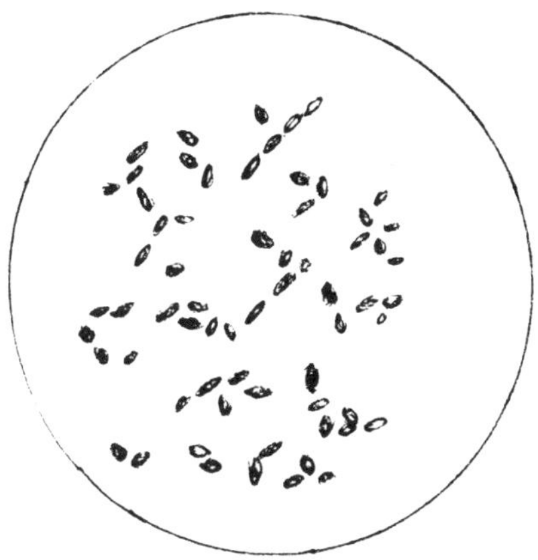

Shigella dysenteriae. The cause of bacillary dysentery.

It is airborne by hungry houseflies, handed out with food in insanitary kitchens and is a regular camp follower of armies. Some antibiotics are effective against it but only good personal care and community hygiene will properly control it. There are many others, — the Amoebae, some harmless some so highly invasive that they eat their way out through the intestinal wall and travel to the lungs, the liver and the brain; the Chromobacteria which can produce bright red pigments and the Pseudomonas which favour fluorescent blue green, bacilli like tetanus and botulisum whose poisonous toxins are the most deadly substances known, but fortunately in this situation remain quite harmless; Proteus and the Coliforms and Paracoliforms, and viruses. Truly the bowel contains a veritable menagerie of living animals.

OVER TO YOU

"To the hush of the breathless morning,
On the thin, tin crackling roofs.
To the haze of the burned back ranges
And the dust of the shoeless hoofs."

The Native Born
Rudyard Kipling

The "Northern Territory" is that part of Northern Australia which lies approximately between 26° and 22° south of the Equator and 129° and 137° East of Greenwich. Its inhabitants cover a range of people from stone age man to the sophisticated technologists of a computerised age, and its geography is decorated with names ranging from the local

Aboriginal Annaburroo and Ooaminaa to those commemorating navigators and scientists from far-away Europe like Darwin, Beagle Bay, Van Dieman's Gulf and the English Company's Islands, and some recent acquisitions like the uranium settlement of Rum Jungle.

Isolation and distance have always been a threat to life when sudden illness comes, and the Territory knows all three.

About 9.30 a.m. one hot December day, the Royal Flying Doctor Service at Cloncurry in North West Queensland, one hundred and fifty miles from the Territory border, received an urgent call. A member of a geological survey party operating about thirty miles from the nearest habitation — the small pub-cum-post office-cum-store of the almost deserted mining camp of Boorroloola had become ill, and the leader of the party tried to call base on his small transceiver.

December in the Territory is the time of heat, storms and static, and even in the best conditions at that time of year, one is lucky to make radio contact at all. At last on their inadequate instrument with only a short range, the party did get Boorroloola and over the crackling air said they expected to have a patient in there by 3 p.m. Would Boorroloola please pass the message on to the R.F.D.S.? So Boorroloola stuttered out a message to Cloncurry 600 miles away, but the December heat tossing the boiling turbulent air over the plains and jungles of Australia's north, caught up the waves and flung them, fragmented, far away from their target. Perhaps the Simpson desert and the Arafura Sea heard the messages from Boorroloola but Cloncurry did not. There was something coming into the transceiver certainly — someone was trying to get someone — but of coherence or intelligent message there was none; just the fierce noise, the unintelligible garble of hopeless spitfire static.

It was then that the hospital matron at Burketown, which on its flat plains near the Gulf of Carpentaria could often receive and transmit when other stations could not, just in, good woman, from feeding her two pet brolgas and always on the watch in difficult weather for emergency calls, picked up the little station unsuccessfully trying to call Cloncurry.

"O.K., this is Burketown. I'll relay. Is that Doomadgee or Mornington? Over."

She gleaned what information she could from what the tropic heat would allow.

"Boorroloola eh? Right, go ahead. An accident you said. No? Thrown from a horse was he? Oh, yes thirty hours ago — long time — better get him in mate. Not what? Not thirty hours ago. Yes, got that alright — thirty miles to go. Wasn't a snake bite eh? Well, I can't hear what his trouble is Boorroloola, but I'll tell the R.F.D.S. you want them by 3 p.m. Over and out."

And so the message goes further.

The Drover Plane memorial at Mount Isa.

"Cloncurry, this is Burketown. Boorroloola's been trying to get you. Eh? Yes, too noisy today, isn't it. They'll have a patient there at 3 p.m. Can't hear what his trouble is but they're bringing him in 30 miles and sounded a bit anxious."

And that's it. Only a third-hand message about a sick man somewhere up there in the 'tiger country'! He might be dead by the time you get the plane into the air or he might not be very sick at all. "Sorry you had to come all this way for nothing, Doc", or "He wouldn't let us send for you till we could see it was too late."

Which would it be?

By quick map reference, the pilot realises that Boorroloola is beyond his range for a direct flight. He must go first to Brunette Downs south of his target, refuel there, fly north to the patient, and he must leave Boorroloola in time to reach Brunette again by last light as he has no night-flying equipment on his Drover.

When it is airborne, the forenoon turbulence catches the little plane droning on at one hundred and eighty miles per hour, tosses it high, drops it earthwards like a stone and buffets it to port and starboard. This is visual flying, and if the pilot mistakes his landmarks he could get lost and exhaust his fuel, so, peering down through the dust and glare, he carefully 'reads' his way across the empty lands. To the

---

FOOTNOTES:

Mt. Isa Mines have lately discovered unusual mineral deposits at Boorroloola and it will be an important future centre.

R.F.D.S. (Royal Flying Doctor Service) of Australia which is funded by both Federal and State Governments and by private donations. This base in north-west Queensland has now been moved from Cloncurry to Mt. Isa and today with more modern aircraft and better electronic equipment difficulties such as those herein described would not occur.

undiscerning eye, this could be dreary country — just vast semi-arid plains with few natural features, and those dwarfed by the view from above them. To the pilot, there are many signposts, but as there is a sameness about them all, they must be seen and marked off in sequence.

Mt. Isa, to the South, has been passed, Camooweal and the border show up right below, all in a haze of dust and smoke, for it is the time to burn off the spear grass and low-protein vegetation in anticipation of the summer rains.

Across the border into the "Territory" and then there's the Georgina. The Georgina River is no river at all — only a series of branching outlines made by the trees which line its course, as they curve about in an intricate pattern telling the pilot that in wet seasons water flows where they grow. So he notes Georgina on his map and looks for other landmarks.

If he doesn't pass directly over the homestead of Alexandria Downs, probably still the world's largest cattle run, the pilot notes the indications, windmills and turkey nests,[1] cattle tracks to water over the dry earth, 'roads' across the plains, smoking 'burnt country' where selected controllable areas have been subjected to the summer fires, perhaps an occasional 'out station', or a mustered mob travelling towards a trucking point.

And so at last, by good observation, careful map reading, compass bearings and experience, the tin roofs of Brunette show up just where they've always been in spite of doubts. It is easy to imagine yourself lost over this country. The monotonous sameness of the flat and almost treeless plain, a slight headwind, a little loss of forward speed due to the vertical journeys caused by turbulence, a little excusable anxiety, almost inevitable when your navigation depends on your own interpretation of the smoky terrain below, lead you to think you are almost there when you are not. When those shiny tin roofs don't appear in the dull brown area ahead where you have mentally placed them, or in the next one you select, or the one after that, the feeling that you are really lost this time can obsess you.

Features seem to change places in the haze. The little stony hill that is Mt. Pleasant by your reckoning, is Mt. Hopeless on your map. That dry creek bed twisting south-west should be Brunette Creek but it isn't or you would see the homestead. Hold on though! That high point just appearing on the horizon could be the one you took a bearing on last time you flew out here, and if so Brunette is just beyond. Then, with remarkable suddenness, everything takes on a familiarity which makes your doubts seem absurd and the distant glint of sun baked roofs

---

[1]elevated man-made 'tanks' fed by windmills pumping from bores.

bearing the welcome name in great white letters shortly confirms your position. There's Brunette!

The Drover and your blood pressure drop down together to the broad red runway where you taxi up to the petrol drums and pump flying miles back into your tanks.

Boorroloola is on the MacArthur River which flows into the Gulf of Carpentaria. There were once some small mining ventures here but they have decayed and with them the bush township which is now the barest of citadels with one hotel selling beer or petrol, groceries or postage stamps, and providing accommodation only in extreme circumstances. The citizenry are mostly Aboriginals, both unskilled and unemployed in the Western sense, except for a few excellent stockmen. Don't get lost here! The pilots call it 'tiger country'. Further south, you could land almost anywhere on the open plain and call up your base; here you'd have to choose between a cliff face and a hardwood forest, but because of the roughness of the terrain, it has more features which make navigation easier than on that flat brown journey to Brunette Downs, and there is less dust haze suspended from the sky in the glittering heat. The pilot sets a course almost due north from Brunette and then picks up the Kilgour and MacArthur Rivers and follows them down: the patient, Brown, is waiting, feverish, toxic and dehydrated, for that thirty miles has been a rough slow journey, painful for the patient, distressing for his companions.

To the doctor, the diagnosis is easy — bacillary dysentery — the kind that met the Japanese in New Guinea. Like him, Brown had seen their agony there — the victorious soldiers of Nippon stopped in their Empire-making tracks by an invisible germ, and the name recalled the filthy disease and the repulsive sights. He could smell again their emaciated shrunken bodies, the fouled surroundings, the shrivelled dehydrated corpses, and was grateful for the means of transport and treatment that would save him from a disgusting death in a lonely outpost.

The hot tea and bush scones in the pub kitchen was a welcome break and the Drover was in the air again in time to reach Brunette Downs before last light.

The turbulence has gone now. The hot unruly air of the day has cooled down to the calm of the approaching tropic night and the Drover appreciates the difference. Nearer to the ground than it was safe to go before, the land marks easily picked up again, it glides smoothly south to the safer plain country.

Should the shadow of the low-flying aircraft fall across a brumby mob, usually a harem of about thirty mares and foals, they gallop off, startled, their magnificent stallion, the perfect leader, circling them and snorting defiance at the unseen intruder. He is their leader because he has fought for them, gaining mastery over every other male. He will be

 The Cunning Seeds

their leader till the day he is beaten by a better, kicked and bitten to death or chased from the scene by a younger stallion when he is too old to defend his territory.

The glare has softened as the afternoon advanced. The shadows which a mid-day sun forbade to appear, now etch the gullies and grow eastwards from the trees. As the sun gets lower, a slate-blue rim, edged with pink, forms from the brightness of the day around the complete horizon, before it too, fades and dispels into night and just before 'last-light' the pilot brings his plane down at a long, low angle to the wide red strip again.

On the runway at Brunette, in the evening coolness, stockmen are exercising and training horses, some of which are captured brumbies. They will work as stockhorses, and perhaps prance on the turf once a year when the 'picnic races' make this station the racing centre of the Territory for a few days, giving social contact and entertainment to graziers hundreds of miles apart and a stimulus to horse breeding for the cattle industry.

With the bright moon still shining and the dawn contesting her monopoly of illumination, it is an easy morning take-off and homing run to Cloncurry base.

A week later Brown, ex-commando, turned geological prospector, lay comfortably convalescent in his hospital bed, wondering how this invisible something the doctor called 'the dysentery bacillus' could have struck him so suddenly down and so nearly kept him there. He recalled the lectures the hygiene sergeant used to give them in the Army and the unconcern or derision with which they were received; flies, fingers, faeces, food; he knew now what the sergeant was trying to save them from. His mother had told him too, about the soldier father he had never seen and Gallipoli, and was it 'enteric fever' they called it then? Brown, the father, had been taken from the afflicted fly-filled peninsula with hundreds of others to die on insanitary, hot, crowded Lemnos. He had met an enemy he could not overcome, the undainty housefly carrying on its contaminated hairy feet Shigella shiga, two microns long and half a micron in diameter, the cause of dysentery.

This cunning invisible foe holds within its small mass, toxins which, when released by the disintegration of the bacterial cell fiercely attack the lining of the bowel wall causing it to swell, inflame, bleed, ulcerate, perhaps perforate, and Brown's tortured body, cramped, purged, dehydrated, fevered, unable to compensate for the heavy loss of fluid, surrendered to this undeclared enemy. He had been well treated according to the knowledge of the day with the sulphates of magnesium and sodium to reduce the swelling of the bowel, iced permanganate of potassium solution to destroy the toxins and hot turpentine fomentations between dry flannel to relieve the distressing abdominal pain.

He had been fed broth, albumen water, arrowroot and sago but he died without improvement. Not this Brown though. It was known now that the treatment given to his father was of little value. Sulphonamides had superseded sulphates, and barley broth had given place to intravenous fluid. And Brown, the son, lived, when Brown the father had died.

At morning tea on the hospital verandah sat Brown the prospector, Jim Kennedy, a ringer from a Gulf cattle station with a leg in plaster, little "Snowy" Daylight, everyone's favourite Aboriginal boy, Charlie Jenkins, the hospital secretary, Big Toby, the Aboriginal head stockman from another Gulf station and the doctor who had just completed his rounds.

"So you met one horse you couldn't ride, eh Jim?"

"Couldn't ride yet, Doc. But I will. The wild colts make the best stock horses when they've been shown who's boss."

"Snowy" who had denied the pain in his appendix till it had ruptured and given everyone an anxious time, grinned broadly at the thought of Jim pitched onto the stockyard rails, and helped himself to another sly handful of biscuits. Charlie Jenkins told Snowy that he was to go down each day now to the R.F.D.S. base for his School of the Air lessons. He'd been missing them since he was in hospital.

"Your wife and the new baby are ready for you, Toby. Going home today?"

"Yes, the Boss has come in for us. When do you want to see Marjorie again, Doc?"

"Two weeks today, Toby unless anything goes wrong earlier. Have you made any arrangements about getting back to Boorroloola, Brown?"

"Charlie has. He's got me a lift with a truck going through to Darwin on Wednesday. Is there anything else you want me to do after I get home, Doc?"

There wasn't. Bacillary dysentery, unlike the type caused by amoebae rarely gives any trouble once the acute phase is over.

The Matron arrived then, carrying the latest Toby — a bundle of newness wrapped in a velvety black skin, with small pinkish mounds for cheeks, chin, nose and forehead, and curling hands and feet which were black on top and pale lemon beneath.

As the Doc went over to the office with the secretary for the daily routine signatures, bachelor Brown helped Toby with his family and a little advice about avoiding bacillary dysentery, feeling that if those microscopic miscreants could so nearly kill him, a tender bud like the new Toby wouldn't have a hope against it.

Which it wouldn't.

He needn't have worried. The hospital staff, the Medical Services, the Maternal and Baby Welfare Clinics, and not least, the boss' wife, had all taught Marjorie and many other native women the proper sterilising routine needed to protect their children, and maybe they had listened more carefully to their tutors than Brown and his mates had to the Army Hygiene Sergeant.

As they drove with the boss down towards the post office, the secretary and the Doc watched them go. "I'd like to get some of those people who talk about racism in Queensland here for a few months, Charlie", the Doc commented, "Do them good eh?"

16

TULAREMIA

You may not be able to catch a rabbit but in much of America and other parts of the world you may catch its Tularaemia, a transmissible disease of which the responsible organism is carried by ticks and lice infesting the cotton tail rabbit which Man, in his constant search for food, has put on the menu. It is the kitchen hand rather than the diner who suffers infection for cooking destroys the bacillus.

The bacillus exhibits one of those novel changes in Nature which probably herald an evolutionary development in that it can infect the larvae of the tick through the egg, giving us a rare example of the hereditary transmission of a bacillus by an insect, though the Rickettsiae have already developed this ability.

It was first seen in ground squirrels in Tulare County in California, from which its name is derived, but has since been recognised throughout most of the world and is often fatal.

The organism which causes it very closely resembles the organisms of plague, but it does not possess that enormous power of proliferation which allows the latter to crowd the cells of the host to the point of death. Like plague, it often attacks the lungs. Its size is only 0.5 microns by 0.2, but though so small it has remarkable power of

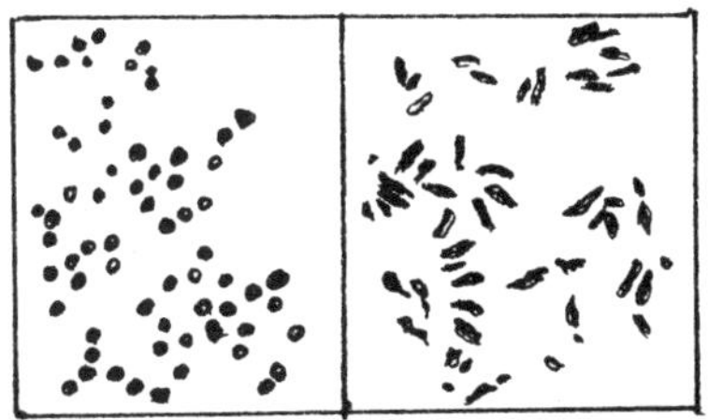

Francisella tularensis showing the coccoidal and the bacillary forms it can assume.

survival, having been maintained in laboratory cultures for twenty-two years, in the skins of animals for over a month, and in the flesh itself for nearly half a year.

Because of its many animal hosts, its efficient vectors, and its powers of survival, it is one of those diseases which will always be with us.

Happily it is susceptible to some antibiotics — notably streptomycin.

COTTON TAILS

O plump head-waiter at the Cock,
To which I most resort,
How goes the time? tis five o'clock.
Go fetch a pint of port!
But let it not be such as that
You set before change-comers,
But such whose father-grape grew fat
On Lusitanian summers.

For this, thou shalt from all thing suck
Marrow of mirth and laughter;
And wheresoeer thou move, Good luck
Shall fling her old shoe after.

But thou wilt never move from hence,
The sphere thy fate allots;
Thy latter days increased with pence
Go down among the pots;
Thou battenest by the greasy gleam

In haunts of hungry sinners
Old boxes, larded with the steam
Of thirty thousand dinners

Will Waterproofs Lyrical Monologue
Tennyson

Giovanni sang as he picked the great black wine grapes on the hot plains near Cassano in his Northern Italy and thought about Maria who had such large dark eyes, shining like the grapes after the rain. But he knew that soon the grape harvest would end and he would have no work and could not marry Maria.

Then he heard Maria coming up the hillside singing as she came trudging barefooted in the poor grey soil of the vineyard, so hot from the midday sun.

She had brought him a letter from Uncle Tony in America and when they read it together, some quick secret tears skipped from Maria's dark eyes to the hot dry earth for Uncle Tony wanted Giovanni to go to America too.

So, when the harvest was over and the grapes had all been trampled into wine in their great vats, Giovanni and Maria, and all their families went to Milano. They climbed the wide pink granite steps which took them from the street to the many platforms of the great railway station, where the trains came in from France, from Germany and Switzerland, from Austria, Turkey, Greece and the Balkans, from Venice, Brindisi, Naples, from the Eternal City itself, to halt briefly under the high steel and glass dome. Everyone wept and laughed, and kissed him on both cheeks. All but Maria who smiled and kissed him and kept her tears till he had gone.

Then the train rushed off to Genoa and Rome with Giovanni and his two brothers to help him to the ship.

Milano! The only city he had ever seen; now running away past him, back to Maria and the vineyards. Would he ever see it again? Milano, the great industrial city of Italy's Northern Plains. Built where the Alpine routes across the Simplon and St. Gothard passes converge to fertile Lombardy. Would he ever be back to eat the grapes and mulberries of these hot, flat lands between the Alps and the Appenines which in past centuries had risen from the encroaching Adriatic Sea and been made fertile by the debris carried down by the River Po and its tributaries from the yearly melting of those high Alpine snows.

These plains, fought for by so many armies. Milano, into which Napoleon had entered in triumph offering liberty, fraternity and equality to a people oppressed by Italian and Austrian alike. Milano, now unable to support all its sons, so that they must seek distant homes.

The plains were not hot and dusty now, for winter and the early snows had come down from the Alps to put a white coat on trees and houses. Suddenly the familiar flatness of Lombardy gave way to the mountains and gorges of Piedmont, which jumped at each other and at the train, as it flicked and twisted around them, roaring in and out of tunnels right up to the Statione Principe at Genoa where it stood still to get its breath awhile before its onward rush to Rome.

It is a long, long platform and as the hurrying crowds go by and disappear, their courage seems to vanish too. Then, with the help of a porter who has bartered his services for a higher fee than he expected, they start in burdened procession towards the Statione Maritima. He is a real midget, this porter, no more than five feet in height. His face is creased in circles by his smile; his skin, dark olive by nature, is made darker by several layers of grease and dust. He wears a grubby cap and a grimy porter's coat, black by design, abuse and age. He would have lived for about half a century, most of it carrying luggage from station, to hotel, to ship, or back again, domiciled in an untidy room close by.

Unlike the high platforms at Milano, these at Genoa are a little below street level and their guide leads them up and out of the wide entrance to the gay Piazza Colombo, so full of flowers in garden beds anchor-shaped in honour of the great Christopher Colombus whose magnificent statue stands amongst them to tell us of the courage of the man who sailed off to discover the new world across unknown seas in three small wooden ships with reluctant crews released from the jails to go on the dangerous journey with him.

Their porter-guide now takes them down a semicircular street, across a small park filled with people and food vendors onto a boulevard along which they follow the curving stone balustrade of the harbour wall right into the Statione Maritima. Amongst the bustle of a departing liner the farewells are repeated and as the ship moves seawards past the long shoal that forms its safe harbour, Giovanni watches Genoa's steep slopes and tiers of ancient grey houses till night and the ocean separate him from Italy, his brothers and Maria.

Uncle Tony had finally settled in a Californian town in the familiar Italian business of restaurant keeper. Childless and needing help, he wanted to establish his brother's son into his beloved business. Giovanni would start right at the beginning peeling potatoes and skinning rabbits for Tony knew that only by doing all the menial tasks himself could the owner know how they should be done.

One day in the fall of the following year, Giovanni, who should have been singing about his work was silent and drowsy, with a headache, pains over all his body, shivers, and a fever. A red spot on his hand grew bigger and ulcerated. He had a pink spotty rash, and hard, tender glands, a constant deep pain in his eyes with ulcers about the lids. His temperature rose and fell. There was inflammation in his lungs and he found it hard to breathe. Giovanni was very ill. Uncle Tony brought the doctor and when he had seen the patient, sat him down with a flask of his best wine.

The doctor explained that Giovanni was suffering from tularaemia, and he thought Giovanni must have caught it from rabbits in the kitchen.

All the great warrior spirits of Italy's past, from Scipio Africanus, who had opposed Hannibal so valiantly on Tony's own native northern plains, to Pompey and Caesar, sprang into the blood of the placid restaurateur and brought him to his feet in an instant. His rabbits! His savoury baked rabbits make anyone sick! His famous Lapin a la chasseur that had made the name of Tony's Restaurant famous throughout California. He almost snatched away the remainder of the flask of wine.

The doctor reassured him. "It is the best meal in California, Tony. You know I come in for it myself."

The light of battle faded from his eyes, and Uncle Tony relaxed.

''Then how do you say the rabbit has made Giovanni sick?''

The doctor replied that it may not be just the infected rabbit, it might be lice, ticks, or flies in the kitchen where the rabbits were skinned.

Uncle Tony leaped again from his chair, clutching his snow-white apron to his heart. Then he sat down again. This poor doctor! He is mad or drunk of course! Lice, ticks and flies in his kitchen! In Uncle Anthony's spotless kitchen! And infected rabbit! When he used only the best of everything!

It took all the old practitioner's diplomacy and another flask of wine to make Uncle Tony understand that even the finest wild rabbits might be infected with Francisella tularensis and that the blood sucking insects that lived on them were not the kind he knew. The larval wood tick already infected from its mother through the egg from which it hatched, would be too small to be noticed even when it bit Giovanni and caused that small red ulcer on his hand. The biting deer-fly which also carried it to man was not the kind found sometimes in less careful kitchens, and the lice were rabbit lice and lived on nothing else. If some of the rabbits he bought, carried the germ of tularaemia, they were still quite healthy and wholesome, and cooking killed out the germs completely. It was only when they were being prepared for the oven that there was a well-known occupational risk in handling them and Giovanni had been unlucky enough to be infected. But Uncle Anthony's kitchen! Ah! there was nothing cleaner in all America.

Uncle Tony prepared to open the third flask when he suddenly remembered his own long days in the kitchen and asked the doctor why he, Tony, who skinned so many of his rabbits himself when he was young, never had this sickness?

As they slowly and slightly incoherently made their way through the last flask of wine, the doctor tried to explain the principles of resistance and immunity, but it was too deep for Tony.

Would the good doctor, please write a letter to send to Italy to Giovanni's Maria?

Maria took her letter where she had always taken her problems — to her parish priest. When he had explained it and her fears had vanished, she went happily back to Momma, at home praying devoutly as she waited for Maria's return. If he were closer, she told Maria, she would run acros to him with some good home food — but what can we do for our Giovanni when he is so far away?

''I can go to California Momma Mia'' she said with a smile. ''I can go to California and marry him.''

17

GAS GANGRENE

The soil of every fertile farm and garden contains a mass of living bacteria in numbers so infinite that we could scarcely devise a unit to count them. They are found almost universally too in the air, on the skin and in the intestinal tract of man and grazing animals.

In their natural home, the soil, they are known as saprophytes engaged in breaking up dead animal and vegetable matter into its elements so that the animal proteins and vegetable cellulose can be restored in simple form to the earth from which they were drawn, to enrich and fertilize it afresh. For this, we all owe them a debt — in fact, we owe our very lives to them, for indirectly they feed us all.

But though normally they keep full the granaries of the world and furnish our daily tables with food, some of them may change readily into the most dangerous of all murderers. It is strange that these organisms, harmless in their natural habitat in the soil, should be so intolerant of a changed environment as to produce in it poisons so vicious that no other substance on earth can match their deadly potency.

There are three groups of these Jekyll and Hyde bacteria which are especially dangerous to man and animals and the condition which triggers this fatal change seems to be an atmosphere devoid of oxygen, for all three produce their toxins only in such an environment.

The gas gangrene group manufacture a battery of toxins which can dissolve the cells of the blood, disintegrate the structure of the flesh, block the activity of the nerve trunks, and ferment the sugars in blood and muscle, setting free gases which react with elements in the body, and the parts become blackened by sulphides and distended with gases. The blackened flesh in which the contained gases crackle to the touch gives the repulsive condition its name, and its untreated end is death.

Less spectacular but just as deadly, are the tetanus organisms.

Unseen and unsuspected, they may manufacture a single toxin, which, attaching itself to cells in the nerve trunks and spinal cord, causes them to send the muscles they activate into the torturing spasms of tetanus from which relief is obtained only in full anaesthesia. One gramme of this nerve toxin purified and crystalised is sufficient to kill six thousand million laboratory mice, if so many could be found.

The toxin of tetanus though, is mild when compared with the poisons which the bacilli of botulism can produce for in the same amount of purified botulism toxin there are twenty thousand million fatal doses for the same little rodents.

And here are these deadly things on the hands of every gardener!

Fortunately though, it is only in oxygen-deficient conditions that they can manufacture their poisons — the decaying flesh in an improperly cleaned wound and the contaminated illegally aborted uterus are common sites for gas gangrene and tetanus infections, and the imperfectly prepared food preserves of the housewife for the toxins of botulism.

It is hard to forgive the obvious tragedies caused by these saprophytes while the good is so difficult to see that it is overlooked, and it is little wonder that they are dreaded. But their iniquities are rare and the blessings they confer are continuous and world-wide. As with mankind, the evil that they do is long remembered and the good so often unknown or forgotten.

Perhaps we should erect monuments to the soil saprophytes instead of laboratories to pry into their private lives and analyse their toxins and their enzymes.

"MY KINGDOM FOR A HORSE"

"How far is St. Helena from the Beresina ice?"
An ill way — a chill way — the ice begins to crack.
But not so far for gentlemen who never took advice.
(When you can't go forward you must e'en come back!)

"How far is St. Helena from the field of Waterloo?
A near way — a clear way — the ship will take you soon.
A pleasant place for gentlemen with little left to do.
(Morning never tries you till the afternoon).

St. Helena Lullaby
Rudyard Kipling

Of all the animals that Man's ingenuity has enslaved, the horse, surely is the most beautiful, the most useful in peace and the most cruelly abused in war. Fossil skeletons show that in his earliest forms he existed on Earth many ages before his master. The ancestors of our modern animal were apparently Oriental, and migrated, probably unaccompanied by man from Central Asia eastwards towards China and Mongolia, westwards to Europe, and south to Asia Minor, India and finally Africa. In Mediterranean regions, they seem to have become involved with prehistoric civilisations which appreciated them for more than their value as food, and though nothing is known about their earliest domestication, they certainly emerged from later civilisations there to evolve finally into our modern breeds.

With the horse to give force and mobility to his warriors, quick escape from his enemies, power to his domestic needs, speed to the hunt and comfort to his travels, Man's village ceased to be his world, and he rode forth to found empires. That is why all early civilisations emerged with the horse, and why the horse and rider became the symbol of mobility and power.

Aggressive man soon saw the superiority mounted warriors would give in battle and consequently when Greek met Persian on land, the Asiatic cavalry often decided the battle. The phalanx formed the main forces of Greece until Philip of Macedonia and his son, Alexander the Great, learned the lessons of attack, mobility, and pursuit possible with cavalry, from having to oppose Asiatic cavalry with European foot soldiers. From then on for nearly fifteen hundred years, cavalry became the main, and in some armies, the only arm of the service, and was the decisive factor in many battles, until the development of gun powder and the bullet forecast its eventual disappearance, which was finally ensured by the use of military aircraft.

We could be certain that so good a soldier as Napoleon would have studied the history of mounted warfare and would use it effectively. He had suffered on the retreat from Moscow from the harrying of the Russian Cossacks who could attack and disappear so quickly.

At Waterloo, after British cavalry brigades had ridden over the French infantry, the Emperor massed eighty squadrons of cavalry against Wellington's centre, but the battered line held against all his dramatic charges. Late on the last weary day of Waterloo, Napoleon sent his infantry forward to attack all along Wellington's thin line, but about 8 p.m., the combined British and Prussian cavalry brigades pierced the French lines which were swept away to be pursued throughout the night towards Paris. Napoleon's short return to his imperial throne from his island prison had cost the lives of 60,000 men at Waterloo.

Nothing comes closer to the popular idea of military glory than the thunderous movement of a cavalry charge — its speed, its irresistible weight, the offensive union of rider and horse, the flashing swords, the levelled lances.

When Lt. Benjamin Miller's brigade galloped into the final assault which cleared the wavering Frenchmen from this dreadful battlefield, the exhilaration of the concerted action removed the thought of personal danger. He scarcely noticed the sword thrust of an adversary till the enemy fled and the battle seemed to be rushing away. His frightened mount, with kicking, screaming, blood soaked horses and groaning men lying or staggering around, and the fury, of which they had all been a part, receding from him, whinnied in terror and started at a gallop after the retreating battle.

The sudden forward gallop, brought Miller heavily from the saddle amongst the lashing hooves of the wounded animals, and as he fell he received a kick which ripped the flesh from his leg, and laid bare the shattered bone. Bleeding, and mercifully unconscious, the young officer lay pale and cold till the bleeding ceased, the body partly restored to his veins the fluid he had lost, and consciousness returned. Night had fallen on the deserted field, where death had quieted most of the wounded, both horse and rider. With bitter pain in his pierced side and battered leg, the subaltern drew a saddle blanket from the dead horse at his side, covered himself with it and slept, swooned, or tossed in pain till morning dawned on the grim and bloody scene; the first morning of 100 years of peace which this fierce battle purchased for Europe, though it had sacrificed two men for every day of that hundred years.

Back in Brussels at last, surgeons pressed the splintered fragments of bone into place, cleansed the mud from his wounds and sutured them, and Miller felt better. Carrier pigeons with a swiftness of communication never known before had taken news of the battle and casualty lists to England, and Benjamin's fiancee was now on her way by coach to Dover and would cross to Ostend with her Mamma and take coach again to Brussels. Father had arranged for merchants in Brussels to supply all his needs and their London doctor had assured them that Sir James Hardacre was the best surgeon in the Army.

Lt. Miller thought himself lucky. He had come out of Waterloo alive and his enemies were now in desperate flight from Blucher's pursuit, and heading towards Paris. The allies would allow them no chance to regroup and plague Europe again and the discredited Corsican would spend the rest of his life in closely guarded isolation.

He began to think of wedding plans when those ugly wounds had healed.

But not all his enemies had fled in disorder from the battle field. Some had entrenched themselves in his damaged leg; punched into the pulpy flesh with the blood-soaked mud of the Flemish farm and they would soon engender a force whose numbers would make the assembled armies of Waterloo look like a raiding party.

Organisms of the group which include tetanus, botulism and gas gangrene are very common in farmland. They form spores, which lie free in the soil, and can remain dormant for many years, till under favourable conditions, bacterial cells will grow from them.

Now in the ragged wound in Miller's leg from which the decaying flesh had not been removed, cut off from the oxygen-carrying blood by the damage to the vessels, the spores of gas gangrene found the anaerobic conditions which exactly suited them, for though, like all living things, they must have oxygen, they will not accept it in the free form but must release it themselves from the material in which they are growing.

In this bloodless and oxygen-starved environment, they absorbed moisture, threw out thread-like arms to become motile bacilli, and each bacillus split into two, the process going on continuously until they were present in multitudes. They manufactured enzymes which liquified the cement of body tissues and caused them to disintegrate; dissolved the red cells of the blood; fermented the sugars and digested the proteins of the damaged flesh, releasing gases which reacted with the iron and the sulphur in the body, producing dark discoloration and the leg became swollen, black and gangrenous.

Sir James ran his fingers over the darkened flesh and felt it crackle beneath his fingers. He pressed a little, and the mark remained. He noted the hot, flushed, restless man, saw the blackness round the wound, the foul smelling bubbly discharges coming from it and knew the officer was doomed. With forced cheerfulness he carried on his rounds and then sat wearily down in his office and sent for Wilson, his assistant surgeon. He looked at the trusting letters which lay on his desk, from the lass in London and her father. He had ordered the standard treatments of vinegar stupes, calomel and port wine. And now mortification! And the boy would die! Mortification? What was it? He knew only that no cases recovered. He might amputate, but the boy would still die.

To Wilson's question about the cause of mortification, Sir James replied that no one knew. It is attributed to uncleanness, gluttony, immoderate drinking and a stagnation of the blood, and of late, even to an infestation of the wound by animalculae which produce poisonous humours. But when he asks to be shown these deadly animalculae, he is told that they are invisible.

In his own opinion, mortification is due to the miasma — the cold fogs of the night. When he served in the Navy he saw little of it, but when men lie out as Miller did on a battlefield through the night mists, it is common. If it is due to animalculae, why do men still die when the limb is amputated. Scientific men do not invoke the invisible to explain their problems, Wilson.

If Sir James was sure Miller would die, Wilson requested permission to amputate the limb high up where the mortification had not extended. Sir James thought of the compliments paid him by the surgeon in London and the young woman now on her way across the Channel. He would give the boy a chance to fight for his own life as he had fought for his country. Wilson, he knew, would operate with skill and dress the wound with commendable care. He agreed.

The high amputation in healthy flesh not yet reached by the organisms removed them completely. The toxins in his circulation were still in insufficient concentration to cause his death; wine and vinegar were kept on the finest linen on the newly cut surfaces so that no other organisms invaded them and Benjamin Miller recovered.

Sir James was delighted. His reasoning was correct after all, and he'd listen to no more nonsense about invisible animalculae in the blood! It was just those miasmata — those stinking mists nowhere worse than in these low Countries!

Oh, to be in England!

18

PLAGUE

Long before ancient navies crippled their enemies by blockading their harbours, the bacillus of plague had used the same process to destroy its unfortunate hosts, for by increasing to incredible numbers, it completely blockaded the infected body cells, cutting off their food and oxygen supply so that they died of starvation. And long before Man became so technically efficient in war that he is able to kill himself off in millions, the swarming cells of Pasteurella pestis, the plague bacillus, restrained his natural increase and kept his numbers within the ability of the primitive agriculture of the day to supply him with food. Tragic as the horrors of the Black Death may seem, the plague bacillus at least killed less viciously than bayonet, gas chamber, or nuclear bomb.

The reservoir of plague is a variety of wild rodents in many parts of the world, but the ordinary house rat and its fleas to which plague is usually ascribed is actually only the unfortunate link between these wild rodents and the human race. When the house rat has become infected, however, it has caused those hideous epidemics of the past, not only because it is very susceptible, and its own rat flea is an efficient carrier of the plague bacillus, but also because once an epidemic commences, mankind's own special fleas and his body lice take up the frightful work and also transmit plague from man to man. And when the horrible little bacillus attacks the lungs to fill them with a plague-derived pneumonia, man can even cough the deadly disease to his own fellows in expelled droplets, without the help of louse, flea or rat.

But let us not place plague and its dreadful massacres in the dim and dirty past. It may emerge at anytime from its hiding places in the cosy burrows of Kurdistani rodents, from the beautiful bodies of Californian ground squirrels, from the gerbils of India, from the field rats of Java, from the prairie dogs of Kansas, from mice in Mesopotamia or Manchuria, from hamsters on the slopes of Southern Russia, from the coasts of Northern Africa, the shores of Madagascar, or the wharves of Singapore, to sow death abroad again.

Though it is a disease of great antiquity and has been described in pathetic detail since man could make records, it is still common in many places and it remains a dangerous disease, ever threatening a too complacent world which imagines itself safe behind a barrier of anti-biotics and health regulations. The now enormous volume of commercial goods and the speed of their transfer about the Earth, including those countries in which plague smoulders dangerously, provide opportunities as never before, for the transfer with them of the plague-infected rat.

What murderous machinery does the bacillus possess that it could kill one hundred thousand citizens within three months in 1348 in the fair city of Florence, or steal across Europe from country to country 300 years later to depopulate towns and villages as it went and destroy twenty-five million people?

This tiny little plump fellow 1.25 microns long and 0.7 across has only one aggressive weapon, a mild type of toxin which increases the permeability of the capillaries allowing it to escape easily from the blood into which the flea has injected it and invade every other part. But it has the remarkable power to proliferate at incredible rates.

From the tiny droplet of blood vomited back by the ravenous fleas, itself blocked by masses of plague bacilli and unable to digest the blood it has already imbibed, this unwelcome colonist multiplies to an astronomical one hundred million in every millilitre of the patient's blood, whose total volume could now possibly harbour more than seven hundred thousand million new inhabitants — 700,000,000,000 using up his oxygen and his nutrients and clogging his vessels and cells. In

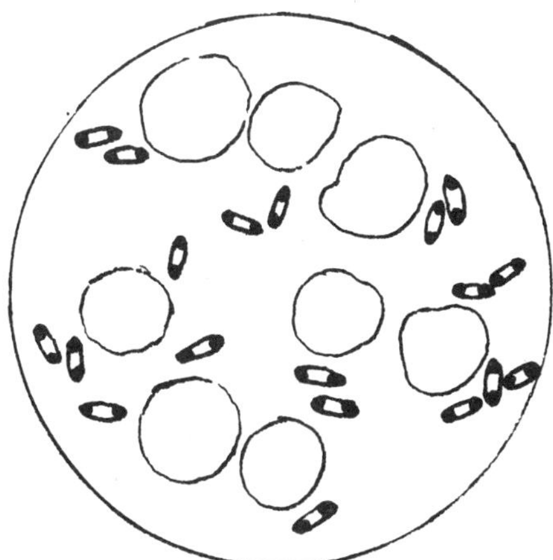

Pasteurella pestis. Organisms which cause plague, showing bipolar appearance.

the other tissues of his disorganised body, they would be present in similar densities — a mass whose numbers would defy estimate — in his swollen tense and tender glands, in his heart, his liver and his brain and in the numerous black haemorrhages formed by blood escaping from his veins. They will crowd the cells and arrest normal activities, and, the plague bacilli and its human victim will perish together.

The plague bacillus is a dangerous unpleasant fellow with nothing at all to recommend him, but he must have some friends or he could not out-live the centuries to renew his slaughter. Some of the wild rodents he attacks, resist and withstand him and then become his permanent reservoir, nourishing and perpetuating him. So long then, as there are wild rodents, there will be plague, and that will be forever. But some of the antibiotic drugs are effective against him and so also is a vaccine prepared from formalin-killed bacilli and containing 2000 million plague bacilli in each dose and so, although he waits on the sidelines still, he will never be able to play his murderous games again as he did in centuries gone. We can do without him.

THE VISITATION

"How can the skin of rat or mouse hold
Anything more than a harmless flea? . . .
The burning plague has taken my household
Why have my Gods afflicted me?
All my kith and kin are deceased,
Though they were good as good could be
I will out and batter the family priest
Because my Gods have afflicted me!"

Medieval Theology
Rudyard Kipling

While the Greeks were growing out of obscurity to be a great sea-power, and devising institutions which would last to this day, their expanding Empire began to touch the fringes of others equally ambitious and warlike, and inevitably the Hellenic world was at last attacked and almost destroyed by the Medes and Persians.

For years the conflict raged between the Greeks and the Persians, until the treacherous defeat at Thermopylae forced both Greek and Spartan to retreat across the isthmus of Corinth, and abandon Athens to their enemies, and it was not until the naval battle of Salamis and the land victory at Plataea finally drove the Asians out of Europe, that the Greeks and the Spartans were able to go back to their own territories.

But when the danger from Persia had passed, the Greek states resumed their former antagonism, their political, commercial, and territorial jealousies.

The Dorian federation of Peloponnesus under the leadership of Sparta, rose against the Ionian federation of the Aegean under the Athenians and with the help of the bacillus of plague eventually destroyed the Empire of the Greeks. With the invasion of Attica by the Peloponnesians in May 431 B.C., the country people crowded into the security of the cities, especially Athens and the conditions there are well described by Thucydides:

> When King Archidemus led the Peloponnesians into Attica the people crowded into Athens finding accommodation in barrels and vultures nests while their enemies ravaged the country side. But while Athens had her ships she could import grain and with the grain she imported plague. It was said to come from Ethiopia and Egypt and it was seen first in Athens in the port of Piraeus, behind the long walls of fortification which led up to Athens, so it was thought that the Peloponnesians had found some way to poison the water in the reservoirs there. Then it appeared in Athens and spread all over Greece. No man knows the cause of it and it defies every remedy. The two sons of my friend Pericles have died of it but though I had it I was fortunate enough to recover.
>
> So many people died of it that the dead lay in heaps in the streets and in the temples where they had gone to supplicate the Gods; but neither prayers nor offerings nor consultations with the Oracles were of any use and at last in despair people gave up all religious observances and all funeral ceremonies and would throw the corpses of their dead on any pyre they could find burning.

And so the Peloponnesian war dragged on with plague the greatest enemy of crowded Athens, for it had much less effect on her scattered foes; internal dissension and political intrigue rotting the structure of this great democracy until in 404 B.C., she surrendered to the Peloponnesian Confederacy under Sparta and saw her Empire dissolved and ''her long walls pulled down to the merry music of flutes''.

If we skip now over nearly two thousand years of history to Florence, we might find Boccacio writing his ''Decameron'' wherein a group of young men and women decided that the only way to escape the plague in their forlorn city was to betake themselves to a remote country area until it was safe to return.

No one tells the story like Boccacio, living through the appalling calamity so similar to the scourge which almost annihilated the Greeks 18 centuries earlier.

> To the cure of these maladies nor counsel of physicians, nor virtue of any medicine appeared to avail or profit aught . . . not only did converse and consortion with the sick give to the sound, infection, but the mere

touching of the clothes, or of whatever other thing had been touched or used of the sick, appeared of itself to communicate the malady to the toucher.

The condition of the common people was pitiable to behold, for that these, in their houses and abiding in their own quarters sickened by the thousand daily and being altogether untended and unsuccoured, died well-nigh all without recourse . . . Nor therefore were the dead honoured with aught of tears or candles or funeral train.

The consecrated ground sufficing not to the burial of the vast multitudes which daily and well-nigh hourly came in crowds to every church . . . there were made vast trenches wherein those who came after were laid by the hundred, and being heaped up therein in layers, as goods are stored aboard a ship were covered with a little earth, till such time as they reached the top of the trench.

So great was the cruelty of heaven that between March and the following July it is believed for certain that upwards of one hundred thousand human beings perished within the walls of the city of Florence . . . Alas how many great palaces, how many goodly houses, how many noble mansions, once full of families, of lords and of ladies, abode empty even to the meanest servant! How many memorable families, how many ample heritages, how many famous fortunes were seen to remain without lawful heirs! How many valiant men, how many fair ladies, how many sprightly youths . . . breakfasted in the morning with their kinsfolk, comrades, and friends, and that night supped with their ancestors in the other world.

And now 300 years later, the phlegmatic Anglo-Saxon meets the danger no better than the amorous Florentine.

Daniel Defoe lived in plague times and with other Europeans trembled at the inexorable approach of the scourage from the Levant. Extracts from his ''Journal of the plague year'' tell the story of one who stayed in London during the terrible year of 1665.

At the beginning of December, 1664, two men died of the plague near the upper end of Drury Lane and ''the family endeavoured to conceal it, but it had gotten some vent in the discourse of the neighbourhood''.

There follows the story of dismay or hope as the deaths increase or decline. They are ''allarmed and unallarmed again'', and the officials alter the Bills of Mortality to hide the true figures until at last ''the infection has spread itself beyond all hopes of abatement''.

It ''grew violent and terrible'' and before it subsided the following year, more than one quarter of the whole population of London had died.

As with Athens during the Pelopennesian War, London had become overcrowded, though with people rejoicing in the Restoration of the Monarchy rather than as refugees from an invading army.

The city and suburbs were prodigiously full of people at the time of this visitation — the numbers of people which, the Wars being over, the Armies disbanded, and the Royal family and the Monarchy being restored, had flocked to London to settle into business, or to depend upon and attend the Court for rewards of services, preferments and the like, was such that the Town was computed to have in it alone, one hundred thousand people more than ever it had before.

The populace became used to the sight of death and as in earlier epidemics seemed at last to care little:

> Sorrow and Sadness sat upon every Face . . . London might well be said to be all in Tears; the Mourners did not go about the Streets indeed, for no-body put on black or made formal dress of Mourning . . . towards the latter end mens hearts were hardened, and Death was so always before their eyes, that they did not much concern themselves for loss of their Friends, expecting that themselves should be summoned the next hour.
>
> There was Desolation and Loneliness and grass grew in the streets of central London.

As in most human tragedies there were some ready to profit from the misery and credulity of others, and the quacks flourished:—

> The posts of Houses and corners of Streets were plastered over with Doctor's bills and papers of ignorant Fellows, quacking and tampering in Physick, and inviting the People to come to them for remedies, which were generally set off with such flourishes as these (viz:)
> INFALLIBLE preventive pills against the plague.
> INCOMPARABLE drink against the plague never found out before.
> NEVER FAILING preservative against the infection.

Throughout the catastrophe the Lord Mayor and the Aldermen of the city of London acted with courage and wisdom. They introduced regulations which, as the real cause of the plague was unknown, omitted to control the only offenders — the rat and the flea. Regulations were published forbidding funeral assemblies, requiring hackney coaches to be well aired and not to be used for one week after conveying an infected person, and stopping all plays, games, bear-baitings and public feastings. All hogs, dogs and tame pigeons, and also unfortunately all cats were to be destroyed, so that by public decree 40,000 dogs and 200,000 cats were killed, because ''they are capable of carrying the Effluvia or infectious steams even in their Furrs and Hair''; — a most informed guess. Some Ratsbane was also laid for rats and ''a prodigious multitude'' of dead rats were found, but it was not realised that they had died of plague and not of Ratsbane. Nor was it realised that those who moved away to the country or lived in ships in the river escaped plague because they escaped the rat and its rat flea.

A few hundred miles away across the water, the Dutchman Van Leewenhoek, was playing with the primitive microscope and was soon to describe bacteria and especially to give the fullest detailed account of the structure of the flea and of its development from egg to adult. It is possible that had England and Holland not been at war the cause of plague may have been found centuries earlier than it was.

Added to the tragedy of death, there was the despair of poverty and unemployment.

> The misery of that time lay upon the Poor, who being infected had neither Food nor Physick; neither Physician or Apothecary to assist them or Nurse to attend them; especially poor Maid-servants who were turned off and left friendless and helpless without Employment, and without Habitation; and this was really a dismal article.

It is hard to imagine the sufferings the little bi-polar, 1.25 micron, Pasteurella pestis has caused the human race. It was the main reason why human populations remained low before its discovery, and have exploded since plague has been controlled, and it is interesting to speculate what different course history might have taken had the observers of the seventeenth century noted even the involvement of the rat and destroyed them instead of the cats. Defoe's informant asks a very pertinent question which really held the key to the mystery and is still unanswered today in respect of many other diseases though not of plague. Having shown the long intervals — seven weeks and nine weeks for example — between cases in the Longacre district he says:—

> Now the question seems to lye thus; where lay the seeds of the infection all this while: How came it to stop so long, and not stop any longer?

And he then concluded that either the disease is not transmitted from body to body or if it is, then a quarantine of forty days only is useless. He was getting near to the point, but being convinced after the beliefs of the day, that the visitation was due to the wrath of God, he gave it as his opinion that the officials had been bribed to report deaths under some other category.

It would have been hard to make people in London in 1665 believe the truth that plague is not really a human disease at all but one of rodents and fleas and that mankind is involved only as an unfortunate third party.

In the year 1894, plague flared again in the Far East and killed some millions of people in northern India and Hong Kong alone, and later spread by ships to many major ports including Sydney in Australia where it is no longer found, and San Francisco in California where it still persists in wild rodents such as squirrels.

It is fortunate for the human race that while some men noisily foment war and destruction, others quietly seek out the secrets of life and death. In 1893, Yersin, a Swiss bacteriologist working in Indo China and Kitasako, a Japanese doctor, independently found the plague bacillus and showed the cause of this baffling disease. Though it will probably always be present among wild rodents in the less accessible parts of the Earth, it is now no mystery and humanity could never again see such epidemics as the past has endured.

Perhaps with time, and the better development of human relations, we may some day be able to say the same of wars.

19

BOVINE BABESIOSIS

Bovine Babesiosis or tick fever is caused by two protozoan parasites of the cattle tick, Babesia Argentina and Babesia bigemina, which by living in the red blood cells of cattle resemble the protozoan parasites of malaria in man. They are round or pear-shaped, about three or four microns in diameter and are carried by the female cattle tick which can transmit them through her eggs to her offspring. When mature, they shatter the red cells, releasing haemoglobin into the blood and when this is filtered out by the kidneys, it colours the urine dark red and gives the common name of "red-water fever" to the condition. The result of continued severe infection is gross anaemia. Acaricidal dips or sprays are used to limit the tick population. The infected animal develops an immunity to babesia which it loses in the complete absence of ticks and therefore of babesia.

When plants or animals are brought into a new environment, they flourish often as they never did in their old habitat. Australia, being still almost a virgin continent when most of the world was already occupied, has suffered seriously in this way at the hands of the thoughtless colonist. The gardener who brought the prickly pear cactus as a desirable succulent for the garden, did not foresee the yearly loss of three thousand acres of arable land to its ravenous advance; nor did the entomologist, who luckily brought the cactoblastus insect to clear it out, envisage such a complete and rapid victory. The Englishmen who imported the fox to provide sport, and the rabbit as a good future source of food, never knew what pests they brought to a new country. The cane toad, which was to devour the destructive cane beetle, has invaded the land in large numbers. And the cattle tick which came from Java on imported cattle with its parasite, the babesia, has cost the cattle industry a yearly bill counted in millions of dollars and a work load which cannot be estimated.

The tiny larval ticks, hatched on the ground, climb up stalks of grass at the urge of some instinct inherited from the dawn of tick life and wait for grazing cattle to whose muzzles or legs they quickly attach themselves. When mature and mated, the female tick needs blood to ripen her eggs and she engorges herself from the host. If her host is infected, she also imbibes with the blood, the babesiae which now, a parasite within a parasite, invade her body and even infect her eggs. Her life cycle completed on the beast, she drops to the ground, and lays her eggs, larvae arising from which will be congenitally infected and will be capable of producing the infection in any new animal from which they in their turn seek blood — for this can be a two-way commerce — the infected animal may infect the clean tick and the infected tick the clean animal. This is a fatal infection and but for the immunity which mild infections produce, few cattle would survive in tick infested countries.

To control babesiosis, the cattle man must maintain the tick in low densities by chemical and other means so as to avoid the blood destruction which heavy infestations produce, but he must not fully eradicate them, for the presence of babesia in the blood induces an immunity, and this will be lost if there are no ticks and no babesia. Then, with a later contact with infected ticks, his stock will be unprotected and may die.

Each female tick lays about 3,000 eggs and so even low numbers can rapidly build up to dangerous populations without careful control. As they do not infest any other animal, cattle are the only source of their very considerable blood requirements.

THE JACKEROO

'Twas merry in the glowing morn, among the gleaming grass,
To wander as we've wandered many a mile,
And blow the cool tobacco cloud, and watch the white wreaths pass,
Sitting loosely in the saddle all the while.

'Twas merry 'mid the blackwoods, when we spied the station roofs,
To wheel the wild scrub cattle at the yard,
With a running fire of stockwhips and a fiery run of hoofs;
Oh! the hardest day was never then too hard.

The Sick Stockrider
Adam Lindsay Gordon

Cattle ticks, a pest imported from Asia, have now spread over all those parts of the continent where the temperature and the humidity

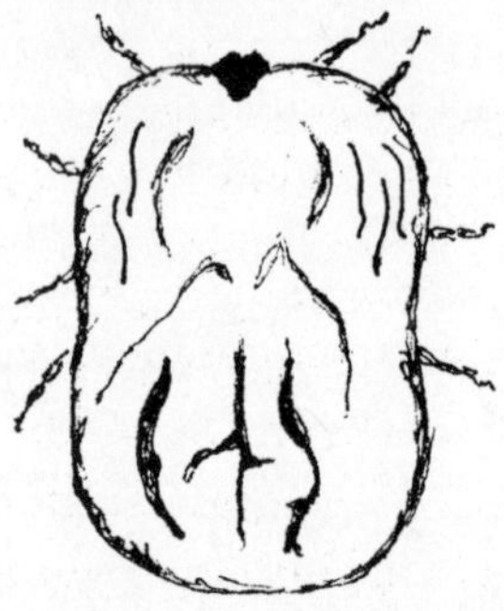

Female cattle tick.

suit them. Blood sucking and highly reproductive, they are also the vectors, the living bridges, by which certain microscopic protozoa — the Babesia, and certain rickettsial parasites — the Anaplasma, cross from one infected beast to another as yet uninfected. The cycle from hatching through larval, nymphal, and adult life till it lays on its own eggs, is about three weeks, and so, to prevent their increase to damaging numbers, the grazier must spray or dip his cattle regularly with effective tickicides within this period.

It was dipping day again on Derryanvil Station and with dew heavy on grass and leaf and just a glimmer of coming daylight, the boss led off, accompanied today by Mr. Scott the Government Veterinarian, who was making his regular check on any resistance which the ticks might be developing to the poisons in current use.

They would follow a routine circuit, mustering by creeks, flats and waterholes on the first day, and sending some of the younger stockmen back with this mob to the homestead while the others went on to more distant parts of the run to converge on the 'outstation' and complete the job there.

The domestic grazing animal, in common with most wild beasts, has something of a 'beat' of its own and an experienced stockman can predict with a certain amount of accuracy, where stock are likely to be found at certain times. He knows they will camp at night on high ground, usually commencing to graze at daylight towards water and shade, if the country is open. This is a good time for mustering and the stockman likes to make an early start. As the day becomes warmer, they are often found lying in resting groups near water and are more difficult to see than the ambulant animal, especially in rough and timbered country. In late afternoon they are grazing again, fanning out across the flats and ridges before nightfall makes it impossible to find or drive them.

The cattle dog is an essential part of the mustering team; in Australia, the favourite is the "blue heeler", an intelligent, tireless dog which, when well trained, will search out the "rogues" hiding in the undergrowth, return a breakaway animal to the centre of the mob, or head off those groups which divide and scatter to avoid the drover. From Cape York to Wilson's Promentory, and from Wilson's Promentory to Wyndham, fractious cattle have learned to respect those sharp teeth in that adroitly-timed snap in the heel which catches them on their weight-bearing foot, even at a gallop.

Somewhere about noon, depending on the heat, the need to rest and water the horses, and perhaps on their own thirst, the men stop to boil up the "quartpots". With one man looking after "the mob", the others stretch out in the thin shade of the eucalypts, refreshed by hot sweet tea and lulled to drowsiness by the continuous song of the cicada, their unsaddled horses hooked by the bridle reins to convenient trees and their dogs panting in the shade and snapping fiercely at the unpleasant flies.

Then the present muster is headed off home, in charge of a couple of "the boys" and the boss divides the rest of the party into two, to muster right and left flanks and to converge down the water courses on to the night camp, strategically placed there for this very purpose. With the vet, he will go there direct to prepare the yards and the "holding" paddock, re-charge the plunge dip to correct strength, repair any break which a startled kangaroo in sudden flight from a predator might have made in the wing fences since last muster, and make preparations for the night's camp.

Besides being the only possible means of transport, the horse is a valuable part of this integrated team. His ears move constantly, each independently to the right or to the left, ahead or backwards, capturing many sounds unheard by his rider; alert to a concealed bunch of cattle in the timber, a few kangaroos camouflaged in the dappled shade, a snake flicking out of sight in the dry grass. When the boss's horse pricked its ears towards a ridge behind which the sun was declining, two fine dingoes appeared right on the crest, a tan male with black trimmings and a common red female, trailing an old and weak cow. Dingoes are no favourites with cattle men though they are intelligent, active, resourceful, cunning, and good hunters, keeping in balance some other potential pests.

The dingo kills for food, of which his choice dish is the newly born calf, often left unguarded at birth by the mother while she goes for water, or hidden during the day as she grazes. He will also attack ill or injured beasts with vicious cruelty, for the snap of his steel-like jaws can mutilate and maim. These two dogs had separated the cow from her little herd, the easier because she had a calf there and preferred to lead them away from it, though she knew they would kill her. Was Captain Oates any braver when he walked out of Scott's tent into an Antarctic blizzard in an attempt to give his companions a better chance of survival?

The dogs will head her up the ridge away from water, each worrying her by turn but keeping out of range of her sharp horns, until, unable to rest day or night and weak from thirst and hunger, she will stumble and fall. Then they will savage her throat before she can rise and bring their pups to find the carcase for themselves, searching across country with sniffing noses held high. They will teach them to find the tastiest parts — to tear the flank and get the kidney — for the female dog with pups to rear is an accomplished tutor and a formidable hunter.

These dogs, intent on their work and silhouetted against the western sky, would have been easy targets but stockman cannot carry rifles on a muster where the timber is thick and the riding hard, though one is kept at the night hut. The beast is recovered, and taken onwards to the evening rendezvous.

The vet volunteered to clear the unwanted fauna from the hut, unroll the swags and make the fire while the boss took the rifle along the creek to get kangaroo meat for the dogs' meal.

The bush population drinks in the morning and at nightfall always with elaborate precautions against waiting predators, so he took up a position overlooking one of the waterholes. A flock of parakeets swished noiselessly into trees near him. A searching pause and one half of them dropped to the water, drank and returned to the trees. The other half did the same, then the whole flock flew screeching and cheering over the water and away down the valley in a green and blue tornado, departing as noisily as they had approached silently, as though with derision at their eluded enemies.

A kookaburra landed heavily on a nearby branch and sat rigid. With cautious reconnaissance he cocked his head to right and to left examining the ground before him and pondering what he saw. With a little hop he reversed his position and studied the rear. Thoughtfully gazing skyward, he raised and dropped his tail several times, nodded and glided down to an old stump near the water. Here he went right through the same performance. Something told him there was an unseen enemy. Something perhaps in the silence and the watching eyes, as a lone human being might feel an uneasy sense of a hidden observer.

He cocked his head to listen, raised the feathers of his neck, lurched backwards and forwards balancing himself with his tail, did several about-turn hops and surveyed the ground, the trees and the sky. Had the scent of a man wafted out into the cool air? Had he noticed an unfamiliar silhouette in the familiar surroundings? In other places he would be untroubled by the presence of a man, but near to water at sunset is dangerous ground. He flew away to keep his thirst till morning.

A tiny honey-eater not much bigger than a large beetle, whipped into the callistemon overhanging the water — a busy little bustling bird, much too busy for caution. He had come to drink but here were blossoms too, and a few insects. He missed nothing. Dropping from the last blossoms, he hung upsidedown from a twig just above the water, drank, and then flashing straight up through the trees with a speed it was difficult to follow, was off and away. Speed and smallness were his defences.

Further down the creek the boss could hear the harsh croakings of a flock of black cockatoos whose black and red beauty and hoarse screeches were as incompatible as pretty faces and vulgar voices. They were feasting on the hard fruits of the banksias and as it grew late they came noisily in above him, heavy and laboured in their flight, but with the gorgeous painted red underneath them in beautiful contrast to their glossy black feathers. They delighted the eye but offended the ear and he was pleased when they had finished their drink and were gone.

With the evening coolness, the kangaroos appeared noiselessly, heads turning to catch every sound; hopping forwards only a few yards at a time, to tempt their hungry enemies to betray their presence, and, even when bending to drink, still ready at the slightest hostile sound or movement to bound away to safety. At the crack of the rifle they fled up the hillside but stopped there again, rigidly erect, inquisitive targets for easy slaughter, while the boss picked up his victim to take back for his working dogs.

Now the cattle are coming in, each mob herded along the wing fences which run out obliquely from the holding paddock, anxious mothers bellowing for calves which, tired by the long trek, are trailing towards the rear, or if very small and weary, may even be carried on the front of the saddles. The gates are closed, the dust settles, the mothers suckle their truant calves, the kookaburras assemble to laugh

Laughing kookaburra.

at the spectacle, the dogs are fed, the horses enclosed in the night paddock and the quart-pots boil again on the log fires.

The cheerful fire of the bush cattle camp is, like death, a great leveller. Each man's possessions there are those simple essentials which can be carried in a roll across the front of his saddle or strapped, like his quart-pot, behind it. His seat is the ground or whatever logs or stumps may be there; his ground sheet and blanket are his bed; and the walls and roof of his abode are the same for all — a circle of light in outer darkness, hung with stars. The differences lie in a man's ability to tell a good yarn, sing a song, or discuss an interesting topic, and many and varied are the subjects thrashed out around bush campfires, when darkness halts the activities of the day.

Tonight it was Harry and Leslie, the new jackaroos, who asked the vet to tell them just how these cattle ticks kill the animals.

It seemed strange Harry said, that cattle will die if they have either too many ticks or not enough ticks, and he challenged the vet to explain this apparent contradiction.

''The boss could tell you all that, Harry'', the vet said, but the boss was busy making more tea and said there was still plenty he'd like to know too.

Seated in this bush schoolroom with his back against a stump and his legs stretched out towards the log fire which occasionally exploded to throw sparks towards the dancing shadows on the rim, the vet began a story it was his delight and partly his duty to tell to the cattlemen he visited.

''Well then, Harry, first of all, cattle are killed by a heavy infestation of ticks simply because they are bled to death. Each female tick takes about two millilitres of blood. Anything up to one thousand ticks is called a moderate infestation in tick research work and of course, neglected animals may carry many thousands. The animal finally becomes so anaemic that it dies.

Harry could understand that part easily enough.

''Now the other way they kill is not because of what they remove from the bloodstream, but what they add to it. This cattle tick is usually infected by parasites, known as babesia, which live within the red blood cells of cattle just as the malaria parasites do in man. The female tick feeds on this infected blood and the parasites, which are released when the red cells are digested, now enter the cells which line the gut of the tick. Here, each babesia multiplies to about 500 and then they migrate to every part. They even enter the ovaries and infect her eggs which can then infect the new ticks which hatch from these eggs.

When she is ready, she drops from the animal, lays about 3,000 eggs and dies, simply because her life-cycle is complete.

The larvae when hatched, climb up stalks of grass and wait for grazing cattle to attach themselves to.''

"John", said the boss, "these babesia must be mighty small if 500 of them can fit inside a cell in the gut of a tick and then go on to live in its eggs".

"They are about 4 microns, that is 4 thousandths of a millimetre in diameter, but that is gigantic compared to, say, some of the small viruses which are as small as one fifty thousandth of a millimetre."

"Astonishing", exclaimed the boss, "and most of us never give such things a thought".

"We all 'step to the music that we hear' I suppose Ted, in our separate jobs; but everyone would benefit from at least some knowledge of the microscopic world."

"But Mr. Scott", Harry said, "what you have told us explains nothing. If your cattle have these parasites in their blood cells and transfer them through the tick and its larvae back to themselves, why should they die now when they didn't die before? If the babesia travel from the cattle to the tick, from the tick to its eggs, from its eggs to its offspring and from its offspring back to the cattle again, where is the fatal difference now?"

The vet smiled. It was good to have enquiring young minds to teach.

"Suppose then, these parasites are injected by the new tick into a beast that has never contacted them before; Do you think it might die, Harry?"

Harry didn't see why it should. "There must be a 'first contact' for every beast. All must have been 'new beasts' once."

The boss chuckled as he got up to stoke the logs and brew still more tea. "Wish you were right, Harry. Jolly good theory. But when you've lived with cattle ticks as long as I have, you'll know what to do even if you don't know why."

The vet told them that in spite of what Harry might think, the babesia do kill many cattle and the boss enthusiastically agreed with him, so the boys asked him to tell them why. He described the parasite and its life cycle entering the red blood cells and multiplying till it burst them, when each new parasite entered a fresh new cell and repeated the destruction, till up to 75% of the cells could be broken up. He told them the purpose of the red cells — to carry oxygen from the lungs to all other body cells and that each red cell contains 300 million molecules of haemoglobin and each molecule can combine with four molecules of oxygen, so that each red cell can transport twelve thousand million molecules of oxygen from the lungs to other body cells. He described the formation of the blood cells in the marrow of the bones at the astonishing rate of two million every second, and their constant destruction after about four months use at the same rate. He told them of the astonishing numbers of red blood cells — thirty million million or 3×10^{12} in a 70 kg man and proportionally more in cattle. "Do you see what wonders there are in living flesh boys?" he asked.

The boys realised the astronomical numbers involved when Mr. Scott told them that at the natural destruction rate of about two million per second in their own bodies, nothing abnormal was noticed, but with the massive destruction which occurred with severe babesiosis, the red pigment of the shattered red cells when filtered off in the urine, coloured it a dark red and gave the infection the common name of "red-water fever".

"I've seen it often", said the boss coming back from the fire and pouring out strong tea into all their quart-pots. "It's exactly like this tea", adding, when they smiled, "in colour I mean".

"I see then, Mr. Scott", said Harry, sipping the dark tea, "the final result for the beast is the same both from the ticks and from the babesia. In one case her blood is taken from her, and in the other it is destroyed within her. But now will you please tell us why it doesn't always kill, and why it has no effect on cattle that have it all the time".

Continuing the instruction, Mr. Scott told them about the fierce interreaction between the intricate destructive processes of micro-organisms and the equally intricate resistant powers of the infected person or animal, and the state of immunity that may result, giving protection because of the "antibodies" present in the blood after recovery.

The antibodies derived from the mother in "ticky" country similarly protected their calves for a length of time, sufficient for them to acquire their own immunity.

Finally, Mr. Scott answered Harry's question about too few ticks, explaining that immunity has a duration which varies with the organism inducing it — sometimes it's for life as with human measles, sometimes for short periods only, as with influenza — that of the babesia lasts about ten months and the animals are then susceptible again. If the numbers of infected ticks on the cattle falls too low, the cattle fail to maintain that latent immunity which keeps them safe. So there is a happy medium between too many and too few ticks in which the cattle flourish unaffected either by excessive blood loss or loss of protective immunity.

"And if they lose their immunity, Sir, what would you do?"

"Give them a shot of blood obtained from an infected laboratory animal, which is standardised to contain enough babesia to rekindle the immunity, but not enough to kill the beast. It contains only 10,000,000 babesia in each dose."

"Ten million!" Harry could have listened all night, but the boss now ordered "lights out".

As there were no lights except the stars, this just meant no further talking and Harry and his mate rolled into their swags determined to resume the discussion with the vet next day.

The bush night holds as much activity as the day. The hunting dingo may scent the cattle but will make a wide circuit from dogs and men. The feeding wallabies coming on the camp, will take to flight, thumping the ground hard with their tails in the first few jumps to warn others. O'possums, koalas, gliders will feed in the eucalypts and many small marsupials and reptiles will scour the ground for food. Night birds, and the fruit and blossom bats, the noisy "flying fox", will fill the air, the boobook owl will call from the trees and the plover from the flats, but none will disturb the resting camp, for of all the sleeping draughts which magic or medicine ever devised, none is more effective than a day spent on a cattle muster.

Starting with dawn, the mobs were drafted through the yards, separating out calves to be branded, males to be emasculated, some to be dehorned, for cattle can be cruel when herded together and horns cause much bruising of marketable beef. These jobs done, they were driven into those yards which lead progressively towards the small 'forcing' yard which opens through a race or 'crush' to the 'dip', a concrete bath filled with tickicide solution. It is forty feet long and about three feet wide, with the bottom sloping from a depth of seven feet at the entrance toward steps at the exit. The cattle plunge into this bath, swim to the exit and walk up the steps into the draining yard where they remain for a short time until their dripping bodies have returned the excess poison to the main bath. Ticks already on the animal are killed in a few days by this and the residual coating discourages access of others for a week or so.

The shouts of the drovers cease and the almost continuous bellowing of anxious mothers for their separated calves settles down when they are all through and back in the bigger yards, free to re-unite.

Then, after the boss has made a final inspection, the gates are opened and they escape with tossing heads and contemptuous kicks, as they run off towards a cooling waterhole after the waterless night and the hot morning in the yards. Dipping day is over again for that part of the station.

The boss and the other stockmen took different routes back to the homestead to check on fences, gates, windmills and missing cattle and Mr. Scott with Harry and Leslie took the shortest and easiest way — the sandy bottomed creeks stringing along between waterholes and billabongs. In the sometimes monotonous eucalypt forest, the watercourses can be so different. Here the tall polished aristocratic gum trees may look down haughtily on their white glistening reflection in the pools. Here the sturdy paper-barks strike a fighting attitude to the yearly floods which have twisted and contorted their golden trunks and boughs. Here too are wattles, common and unnoticed in the crowd until they burst into yellow bloom, and here the dainty bottle-brush trees bend over to kiss the water they love so much. Standing a little further back, the casuarinas sigh in the breeze because they are crowded away by the others and have had to grow high and thin to be seen at all.

The bloodwoods, with red gum exuding in congealing clots where their thick bark has been injured; the messmates with the friendly name, tough and useful, if not handsome; the stringybarks whose fibrous dry coat is always ready to light your fire or whose dense yellow wood will serve man's purposes for a century; the 'spotted' gums, blotchy with grey and pink, whose crushed leaves will fill the air with a fragrant perfume; the iron-barks, steel-grey, ridgy, resistant to time and weather, but not resistant enough to defeat the termites munching jaws; these do not grow usually by the waterholes and creeks but on the hard dry ridges; while the brigalow, the coolibah, the mahogany, and the bellbowrie love the moist alluvial flats.

The creek beds between waterholes were often small gorges, narrow and rocky, but cool and shady; impassable sometimes because the stems of stunted callistemons, tilted forwards by the surging waters of many wet seasons, presented a phalanx of opposing lances; sometimes though, easily followed along the well-worn pads of generations of thirsting cattle.

In more open country the boys rode up beside Mr. Scott to ask him about that 10,000,000 parasite dose. How could they be seen or counted, for example?

"Easily enough. By calibrated dilution and microscopic counting of a measured amount. When such a diluted solution is put under a cover-slip and examined by a high power microscope, each field examined contains about a one millionth part of a millilitre so you count the parasites in a series of fields and multiply accordingly. There is nothing difficult about these microscopic techniques for the examination of living matter, boys. What *is* amazing is the living matter itself — its terrific complexity and organisation. This parasite which you think almost unreal because it is small enough to inhabit the egg of a tick, is itself made up of millions of cells and in each cell are the chromosomes and genes of inheritance."

The boys reminded him that they were expected to know something about genes in their study of cattle breeding.

"Well", said Mr. Scott, "here we are nearly home. If you young men will hose and rub my horse down and feed him, I'll have a hot shower to take the stiffness of a day in the saddle out of my not-so-young body and will tell you a bit more after dinner tonight."

"Why is it", asked the boss, pouring his second cup of coffee, "that no drink ever seems quite so good as that bush tea boiled up in our old black quart pots round a log fire?"

"You don't think perhaps imagination helps a bit, do you?" said the vet settling into an easy chair. "But if we start discussing the merits of bush tea with a bushman, Harry, there'll be no time left to talk about ticks. Did you boys look after my horse?"

Of course they did. Now he would keep his part of the agreement.

In the comfort of the homestead lounge in contrast to the rough conditions of last night in the bush, he told them the story as far as was known of heredity — of the double-stranded genes within the chromosomes of the cells, and the genetic code within the genes, and all the complexities of inheritance and reproduction. He showed them why Brahmin cattle whose genes had learned to resist cattle ticks by centuries of contact, were now being so widely used to give an infusion of Brahmin genes to British breeds of cattle which had not developed these powers. It was a long and complicated story.

Well, it was bedtime. Mr. Scott had another journey tomorrow, back to the city and the boss had another paddock to muster. The boys had some interesting facts to ponder and realised that beneath those red and white hides there was amazing activity, invisible and unheard. No wonder cattle breeding was interesting, with millions and millions of genes working for you, and body cells doing their bidding. And no wonder it was necessary to build on the very best foundation cattle you could get, for good points took a long time to breed in, and faults a longer time to breed out. Now they knew why the boss "culled" so heavily.

"Keep a faulty breeder", he always said, "and you will find the same faults in its progeny years later. But breed for good points and you will find the quality steadily but surely improving".

20

BRUCELLOSIS

Brucellosis is a septicaemia of man and animals produced by small organisms which prefer certain animals as hosts but are easily transferable from one to the other and also to man, in whom they produce first a remittent or undulant type of fever which may go on to a long distressing chronic phase lasting for years and often ending in death. They are 0.4 to 0.8 microns in diameter and only 0.4 to 3.0 microns in length so that their shape is roughly coccoid or coccobacillary. There are three strains and like Tweedledum and Tweedledee they are so close in size and appearance, and also so identical in such characteristics as motility, spore formation, and serological responses that they can be identified only by special techniques. Fortunately they are rather colour sensitive and react differently to various dyes and they have slightly different needs of oxygen so it is just possible with great care to know them apart. Though they were at first studied independently as separate organisms they are now recognised as three strains of the one bacillus — the caprine, a parasite of goats named Brucella meletensis, the bovine in cattle named Brucella abortus, and the porcine in swine named Brucella suis. They produce a contagious type of abortion which in the United States alone is estimated to cost the livestock industry one hundred million dollars yearly. They are transferred to man by the drinking of raw milk or by eating cheese prepared from infected animals, or by contact with the contaminated discharges from aborted animals. Once thought to be confined to Mediterranean countries where it was given the names Malta or Mediterranean or Undulant fever, it has now been found to be a world-wide scourge and has become an increasingly grave public health problem, because besides the risks through infected goats, cattle and pigs, it has been found also in sheep, deer, buffalos, horses and dogs. Many countries including

Australia are now attempting to eradicate it from their domestic animals. This insignificant tiny organism approximately half a micron in diameter which does not even seem to know what shape it wants to be, whether a coccus or a bacillus or something in between, has had dramatic effects on the wars of mighty empires. It has affected the trade and commerce and even the politics of nations at the opposite ends of the Earth. Like the babesia which parasites cattle through the tick it has added a great load of work to farmers and veterinarians throughout the world and because it also afflicts human beings, to their doctors also. On entering the body they travel along the lymphatic circulatory system and enter the blood which carries them to liver, spleen, bone marrow and defensive cells which increase enormously in an effort to destroy them and in all these organs they may live, proliferate and produce serious symptoms of illness even for years.

In 1844 almost fifty years after the great naval battle which led to British possession of Malta and the Ionian islands and gave her a valuable Mediterranean base and protection for her interests in the East, there was raised in Trafalgar Square in London the magnificent granite monument known as Nelsons Column, one hundred and seventy feet high and decorated at its base by brass plates made from captured French cannon, with the little Admiral standing high atop above the busy streets. It was the nations grateful tribute to her greatest sailor by whose naval genius allied to the equal military skills of Wellington, the bitterness of Russian winters and especially the steadfast courage of the beleaguered British people over a quarter of a century, Napoleon was finally banished to St. Helena.

The hosts from all parts of the world which crowd this famous Square would see little relationship between Nelsons Column and sick hogs in Chicago, goats milk in the Mediterranean, and aborting cattle in Australia, which, in 1805 when Trafalgar was fought was almost unknown. Yet the fevers so common on the islands and the shores of the Mediterranean which were a serious military threat to the garrisons both French and English stationed there, helped Nelson's blockade and assisted in expelling the French. It was the military necessity to defeat this invisible foe that led to its eventual discovery and indirectly to todays efforts around the world to eradicate it entirely from our domestic animals.

Malta less than eight miles across and twenty miles long seems to have been born in calamity and to have lived with wars. This small group of islands, only 180 miles from the African coast and sixty miles from Sicily are all that remain of the tertiary limestone land bridge which once connected Africa and Europe. Malta is divided into two levels by a great geological fracture which runs diagonally across it and the other islands indicating the violence of the Earth's movements which separated the two continents, and cut off elephants and rhinoceros from their African homeland leaving them to die and deposit their fossil bones in Malta's limestone caves.

Stopping short of complete submergence this terrestrial subsidence left Malta with its one great asset, its magnificent harbour — its delight and its disaster, its fortune and its misfortune. Here the mariners of prehistoric civilisations sheltered their little ships, took over the islands and left some of their history in imposing ruins. Peoples of Mediterranean origin, men of the Stone Age, Phoenicians of the Iron Age, men who had learned to work in brass, the Romans, the Carthagenians, the Sicilians, the Knights of St. John of Jerusalem expelled from the Holy Land and island of Rhodes by the Arabs, sailed their ships into this fine harbour and left their marks on the little island. It was attacked by the Turks and the Aragonese; it was taken over by Napoleon; it was blockaded by Nelson. It revolted against the French masters, and assisted by famine, Malta fever, and British troops expelled them and became a British naval base when eventually Napoleon ceased to trouble Europe. It was made wealthy by its strategic and commercial importance during the Crimean War for it was at Malta that the first detachment of British troops put in on their way to the Crimea before going on to another place made famous by war — Gallipoli. When war outgrew its local significance and became global and not just a Mediterranean affair, and the wooden trireme gave way to the iron-clad battleship, Malta and its lovely harbour became a base for great navies and also a hospital for the sick and wounded of Gallipoli and Salonika. Again when mankind sought to attempt its own destruction in a second World War it became a daily target for Hitler's bombers based in Sicily only twenty minutes flying time away. For its courage and suffering the little island was awarded a unique honour in 1942 — the George Cross.

Captain Marston sent out from England towards the end of the Crimean War to help in the very inadequate medical care of British troops in that catastrophic campaign, was posted in 1856 as medical officer to the garrison at Malta. Life was pleasant here — a delightful climate in place of Russian blizzards; peace and plenty instead of Crimean suffering and war; but there was one drawback — this horrible fever that attacked so many. Marston had seen plenty of illness in the Crimea — dysentery, cholera, malaria — but this was an insidious fever which came on slowly and kept returning, and it made men so weak and listless they were unable to work at all. They might suddenly fall ill with sweating fever, chill and headache, and severe body and chest pains, but usually they would have a feeling of ill health and low remittent fever for weeks or months before reporting sick; and Captain Marston was beaten. All he tried to do for them and all his rules about cleanliness and flies were quite useless. Even the wives, some of them models of all the new teachings of Florence Nightingale, got sick too.

Then one day in 1859 Marston got it too. He decided to keep a complete case history of himself and for the first time the doctors were given a full and reliable account of it all, for by the time he was fit to resume duty he had personal knowledge of the troublesome fever, the sweats, the chills, the weakness, the joint pains, the muscle pains, the tender liver, and the wretched headaches. His observations were published in 1863 but though it was a start it didn't show any real cause of the malady. At least it relieved the harassed garrisons from the charges of laziness and malingering of which they had so often stood accused.

Our knowledge is usually gained slowly and painfully, and the astute observations of one worker may lead another to important discoveries many years later. Poor Captain Marston is not heard of again, and the honour of discovering the cause of his Malta fever came to Major David Bruce in 1887, when he was able to isolate and grow an organism which he called Micrococcus Meletensis, from the spleen of a patient who had died of the disease. The name was later changed to Brucella meletensis in his honour.

In 1897 two other investigators Wright and Semple developed an agglutination test which put future investigations on a scientific basis and the name undulant fever was suggested by Hughes because of the remittent nature of the fever it induced.

Also in 1897, Bang working in Copenhagen found a similar organism which was causing abortion among cattle and showed that it localised in the udders of pregnant cows. He called it Bacillus abortus, which, now that its relationship to Brucella meletensis has been shown has been changed to Brucella abortus.

So far no one had shown how to treat the disease successfully, so in 1905 a Mediterranean Fever Commission was set up and one of the workers in that Commission, Zammit, found that the serum of 5 of 6 experimental goats gave strong positive agglutination tests for Malta fever. Carrying the investigation further Horrocks found strong positive tests in 40% of goats, and, that their milk carried the organism.

Eighteen years had gone by since Bruce identified the organisms in 1887 until Horrocks demonstrated its presence in goats milk in 1905. The next year, 1906, an Army order was issued forbidding the use of raw milk and there was an immediately dramatic fall in the incidence of Malta fever.

Investigation did not end there. In 1914 Traum in America discovered the third member of the group, Brucella suis, in hogs and in the same year Kennedy found agglutinins to Brucella abortus in the blood of cows near London. In 1918 Alice Evans clearly demonstrated that they were not separate bacelli but that Brucella meletensis, Br. suis, and Br. abortus were all strains of the same bacillus.

Throughout the nineteen twenties this little organism was identified in various countries in France, Italy, Algeria, Rhodesia, Palestine, Canada, England and Wales, United States of America and in European countries so that it exists in most parts of the world. In the century since Sir David Bruce first discovered this organism it has been closely examined by many laboratories in many countries and still all is not known about it but at least we can control it and perhaps in time it may be possible to banish it from our flocks and herds.

This is a good example of successful scientific co-operation by many countries and of the need to study the animals of the world and their ills as well as our own for these are often interrelated.

21

VIRUSES AND CANCER

"And does the road wind uphill all the way?"
"Yes, to the very end."
"Will the days journey take the whole long day?"
"From morn till night my friend."

Christina Rossetti.

Research, or the diligent enquiry after new knowledge has become the province of teams of specialised workers in various disciplines using highly sophisticated and expensive equipment, and now rarely relies on the work of the individual researcher. In the field of Microbiology, the Microbiologist teams up with the Physiologist, the Mathematician, the Physicist, the Geneticist, the Molecular Biologist, the Pharmacologist and the Biochemist, and by no means less important, the Statistician the Technician and the Librarian. Little wonder then that the road leads into strange and interwoven regions of knowledge of which even the interpretation can take years.

It has been known since 1908 when viruses were first shown by Ellerman and Bang to produce cancer in chickens, that some viruses can cause some types of cancer, and since this connection was reported the road has led over three quarters of a century with notable milestones along the way. Rous discovered sarcoma in poultry in 1911; Shope, myxomatosis of rabbits in 1932; Bittner, mammary gland tumour of mice in 1936; and Gross, leukaemia of mice in 1951, and other examples of virus-related cancer have been found in cats, turkeys, frogs, squirrels, cattle, deer, monkeys, guinea pigs, hamsters, baboons, gibbons, horses and rats, and such probably exist in every other vertebrate. Even the mice that inspired Bobby Burns to compare them favourably with his own equally lowly estate, suffer, it seems similarly with men. So in animals oncogenic or cancer producing viruses, ranging mostly in size from 40 to 100 millimicrons occur naturally and can produce a variety of conditions including leukaemia, lymphoma, sarcoma, papilloma, and breast cancer, and though they are readily transmissible in the laboratory to others of the same species there is no experimental proof that they can be transmitted to human beings.

Can we then consider ourselves, perhaps superior in some unknown way to those ''Earth born companions and fellow mortals'' in which viruses can be made to induce malignant neoplastic disease? Or since, when they do affect human hosts they produce diseases apparently identical with those in lower animals, can we conclude that there is, perhaps, no great difference in all Earth's creatures, including mankind, after all, and that those things which cause disease in the one will produce the same affliction in the other?

Only two tumour-like human conditions are known to be caused by viruses — Warts, and a similarly warty and contagious malady, known as Molluscum contagiosum. Neither of these threatens life. Why do they show some of the characteristics of tumour and still remain benign? What is the link between the malignant and the non malignant? And why cannot virus-related cancer in animals be transmitted to humans as it is to other animals?

Today, with the welcome help of the electron microscope in the study of the sub-microscopic viruses, with new knowledge of the mechanisms of heredity, with many technological advances in the laboratory, and with great progress in the field of cellular biology, research into the causes of cancer is going rapidly forward and giving new hope of eventual victory.

Someday perhaps, at the end of the long, long journey, when researchers have unravelled the tangled skein of the groups of Oncornaviruses, Papovaviruses, Adenoviruses, Herpesviruses, Cytomegaloviruses, and all the other viruses and virus prototypes; when the naked virions and the membrane-bound virions have lost their mystery; when we know more of the hereditary machinery of the cell, and clearly understand all about the double helix of the double stranded chromosomes of heredity; when we know why some viruses will induce cancer and some will not, and whether they act alone or need the aid of other physical or chemical factors like radiation, or tobacco smoke, or hormones, or a combination of them; when we know what role healthy or defective genes play in the processes; we may be getting near to knowing what causes cancer and how it may be prevented.

There has been progress. Present research indicates, but by no means proves, that mouse mammary adenocarcinoma and human breast cancer may have a similar virus origin. Experimentally, infected mice have been shown to transmit leukaemia virus in their milk and so infect their offspring.

It has been possible to vaccinate poultry successfully against a malignant and highly infectious disease due to a Herpesvirus, — Mareks disease — and it seems that it should be possible to protect humans in the same way against other Herpesviruses.

A virus-related cancer of the eye, which affects cattle in Queensland is showing promising signs of cure by immunological means. There is also a environmental factor here, because it is known to be at least partly a result of strong sunlight, and possibly there is a genetic factor also, in that some English breeds of cattle are susceptible while the Brahman types are resistant. Could the Brahman, arising in the tropics, possess a resistance to this environmental factor which, the milder European climate has not induced in the British breeds, or is there something genetically different which protects the one rather than the other.

The dangerous germs of diphtheria which have killed so many children appears to be harmless until it is associated with a virus. Is it the virus that kills or is it the deadly partnership of bacillus and virus?

Maybe the research scientist is now on the way towards finding protection for mankind against this often silent and invidious foe, at least in some of its manifestations.

But it will be a long and uphill road still.

It will need the best brains of the best teams every country can put together, and it will need more than brains and dedication and devoted tedious work. It will need money. Money from the small savings of ordinary men and the private fortunes of the rich, and grants from National and International budgets; for modern methods of investigating the means of preserving life are, like the modern means of destroying it in war, immensely expensive.

This is a war for all mankind, — the only war in which there can be no losers.

Glossary

AEROBIC: In relation to bacteria, aerobic means having the ability to live in the presence of air, and to use atmospheric oxygen.

AMINOACIDS: The sub-units of proteins. Bacteria in common with other living things, require aminoacids and many can synthesise their needs themselves. The gram-negative organisms are much more efficient in this than the gram-positive organisms. Salmonella typhi for example can make all except tryptophan.

AMOEBAE: Single celled protozoa which move by protruding a portion of their bodies and drawing themselves forward with a sort of flowing motion. Though most of the amoebae are non-pathogenic, some produce serious human diseases.

AMNIOTIC FLUID: The fluid contained in the membrane (amnion) in which the foetus or embryo of mammals, birds and reptiles is enclosed. The embryo or foetus thus floats in the fluid which protects it against injury. Amniotic fluid is used to grow various micro-organisms in the laboratory.

ANAEMIA: Broadly, anaemia means a deficiency in numbers or quality of the red cells of the blood.

ANAEROBIC: In bacteriology, denotes the state in which micro-organisms are able to live and grow in the absence of free oxygen. Such organisms obtain their oxygen by breaking down chemical compounds containing oxygen.

ANGSTROM UNITS: (Symbol A°) The unit used in measuring the wave length of light. $1 \text{ A}° = 10^{-.10}$ metres = one hundred millionth of a centimetre.

ANOXIC: Having insufficient oxygen supply.

ANOPHELINE: (From Greek anopheles = harmful) A genus of mosquito which is the most active in the transmission of malaria. It has dappled wings and when drawing blood it adopts an upright position at an acute angle to the skin.

ANTIBODY: Antibodies are immunoglobulin molecules secreted in the tissue fluids by the lymphoid cells in response to exposure to foreign substances which are known as antigens (q.v.). Antibodies are produced particularly to invading bacteria and viruses, but they may be developed in response to other substances, often with less fortunate results. They may, for example, cause allergic effects, the rejection of implanted tissues, the agglutination of incompatible blood in a transfusion, or sometimes the destruction of a new born baby's blood when the parents blood is of a different Rh group. Antibodies are highly specific for the particular antigen which causes their production and act only with that antigen.

ANTIGEN: Antigens are protein substances which when introduced into the body either naturally or artificially, are recognised by the defensive cells as foreign substances to be neutralised or destroyed. These defensive cells, thus stimulated, produce antibodies which act in a variety of ways. They may precipitate the antigen from solution (precipitins); dissolve them, (lysins); agglutinate them into non-functioning clumps (agglutinins); neutralise them, (antitoxins and antivenines) or they may cover the surface of the antigen with a substance which is not repellent to the phagocytes and so assist phagocytosis. All immunising techniques depend on these principles.

ANTIGENIC: Capable of acting as an antigen and so producing antibodies.

ANTITOXINS: Substances capable of neutralising toxins, such as the toxins of diphtheria or tetanus or snake venom (antivenine).

AORTIC INCOMPETENCE: The aorta is the main artery of the body. It arises from the chief pumping chamber of the heart, the left ventricle and distributes blood to most parts of the body. Its inlet is protected by strong valves which prevent the return of blood to the heart as it relaxes to fill again for the next beat. Irritation and inflammation to the aorta and these valves cause damage, resulting in incompetence of action and because the spirochaete of syphilis commonly attacks these structures it is a common cause of the condition.

BACTERIOPHAGES: Usually shortened to 'phages. Ultramicroscopic organisms probably viral in type and form 10 to 65 millimicrons in diameter, virtually cause virus diseases of bacteria, destroying them. As they are strictly specific for each species of bacteria they provide an additional means of identification by what is known as "phage-typing".

BILLABONG: A lagoon which remains isolated in an old watercourse because the stream has taken a new direction.

BIPOLAR STAINING: Some organisms take stains at either end but not in the centre as e.g. the plague bacillus.

BOORROLOOLA: Once a small deserted mining village in the Northern Territory of Australia, it has now been re-established as an important mining centre by Mt. Isa Mines.

BROLGA: A tall stork-like bird common in rural areas in Australia. It is remarkable for the great beauty of its mating dance.

BRUMBIES: Wild horses, the progeny of station animals which have escaped from domestication.

BRUCELLOSIS: Sir David Bruce, an army officer stationed at Malta, first isolated the organisms of undulant fever and the Genus is known as Brucella. It contains three species, Br. abortus, Br. melitensis, and Br. suis which cause brucellosis in cattle, goats and pigs respectively, and can be transmitted to man.

BUTYRIC ACID BACTERIA: Bacteria which promote the formation of butyric acid as in butter.

CHROMATIN: The protoplasm of the cell nucleus.

CHROMOSOMES: Threadlike structures in the cell nucleus which contain the genes which are the physical basis of heredity. When the cell is about to divide the chromosomes undergo a complicated rearrangement which ensures that the new daughter-cells will contain an equal number of chromosomes and therefore of genes.

CICADA: (cicada — cricket L.) Tree crickets of the order Cicada. Their loud humming noise is produced by the action of their transparent wings.

CILIA: Tiny hair cells which line some cavities such as the respiratory passages and help to retard the ingress of foreign materials.

COCCI: The shapes of bacteria are more or less of five types — the cocci are spherical, the bacteria rod-shaped, the vibro the shape of a comma, the spirochaetes take a spiral form and cocci-bacilli are intermediate. There are many variants of these general forms due to spores, flagellae etc., and in the aggregate they may form chains, bunches, bundles or geometrical packages.

COTYLEDONS: The parts of the placenta which nourishes the growing foetus as the seed-leaf or cotyledon nourishes the embryo plant.

COVER-SLIP: A fine circle of glass used to cover a specimen in readiness for microscopic examination.

CYTOPLASM: The protoplasm of the body of a living cell surrounding the nucleus.

DARK-FIELD ILLUMINATION: A method of illuminating objects for microscopic examination by transmitting light across the field instead of directly through the condenser. Any objects in the field are thus tangentially lighted and appear very bright against a dark background.

DINGO: The native dog of Australia — the warrigal.

DIPLOCOCCI: Cocci which appear to grow in pairs sometimes more or less kidney-shaped as the Neisseria and Meningococci.

D.N.A.: Deoxyribonucleic acid. The chemical in the cell nucleus which controls the genes in the chromosomes both in respect of their function and their replication.

DOUBLE HELIX: Double spiral. This is the form which is assumed by both D.N.A. and R.N.A. (q.v.)

ELECTUARIES: A pleasant tasting disguise for medicinal drugs made into a paste with honey, treacle or syrup. Rarely seen now.

EMBRYO TISSUE: The substance of a growing human being in the first three months of intrauterine life, or of the animals in various uterine stages.

ENDOCARDITIS: Inflammation of the endocardial membrane which lines the cavities of the heart.

ENDOTHELIAL CELLS: Those cells lining the inside of such structures as blood vessels in contrast to epithelium which is external, e.g. on the skin. Special types exist, e.g. germinal epithelium, q.v.

ENDOTOXIN:
EXOTOXIN: See toxins

ENZYMES: Biological catalysts which produce very rapid chemical changes in organic tissues.

ERYTHROCYTES: Red cells of the blood (q.v.) erythros = red (Gk) cyt = a cell (GK)

ESCHAR: The dry scab which follows a burn. The same sort of scab often follows the bite of a mite and is commonly seen in the scrub typhus patient.

FALLOPIAN TUBES: The tubes in the female which receive and conduct the expelled ovum from the peritoneal cavity of the abdomen to the uterus.

FIMBRAE: A fringe or fine border. Here it refers to the fine waving fringe or finger-like tissues which surround the opening of the Fallopian tubes and by their waving action create a current which draws the ovum to them.

FLAGELLUM: Fine thread-like projections, single or multiple, which by their waving action provide motility for some organisms.

FOETUS: The developing unborn offspring after the embryonic stage. In human beings after the first three months.

GAMETES: Reproductive cells which by uniting, form the beginnings of a new individual.

GAMETOCYTES: The male and female cells which are formed by a deviation from the usual sequence of trophozoite to schizont in the asexual life cycle of the malarial parasite.

GAMMA GLOBULIN: By separation and analysis of the proteins in blood serum it is found that antibodies if present are always attached to the gamma globulin fraction of the proteins. It is in fact, fairly certain that the serum gamma globulin molecules are the antibodies and it is these that are used to confer temporary immunity on such people as pregnant women if they have never had rubella but have been exposed to the risk of it during the first three months of pregnancy or similarly to protect small babies against measles in the presence of an epidemic.

GANGRENE: Death and decay of a part due to lack of oxygen often because of damage to the blood supply.

GENES: The units contained in the chromosomes in the cell nucleus, by which hereditary characters are transmitted to the offspring.

GERMINAL EPITHELIUM: The membranes from which spermatozoa and ova are derived, found lining the seminiferous tubules in the testes in the male and covering the ovaries in the female.

G.P.I.: An abbreviation of General Paralysis of the Insane used to describe the clinical condition of the advanced neuro-syphilitic patient.

GRAM STAINING: Gram was a Danish physician who in 1884 devised a method of staining bacteria for microscopic examination which has become the most important in bacteriology. It depends on the power of an organism to retain aniline dyes of the gentian violet type when subjected to the decolorising effect of alcohol. Those that retain the stain are known as gram-positive and those that give up the dye to the alcohol and so lose their violet colour are known as gram-negative.

GUMMATA: (Plural) The ulcers of syphilis which are often found on the lips and tongue and in almost any organ in the body.

HAEMOGLOBIN: The oxygen-carrying red pigment in the erythrocytes of the blood.

HAEMOLYSINS: Substances which break up blood cells. Many bacterial toxins and some snake venoms work in this way.

HAEMOLYTIC: Capable of breaking up blood cells.

HAEMOPHILUS INFLUENZAE: This is a misnomer as the organism is not the cause of influenza. It was so named because it was frequently found in influenza cases while the virus which is the real cause escaped detection because it was ultramicroscopic.

HISTAMINE: A toxic substance liberated from damaged tissues. It causes increased dilation and increased permeability of the capilliaries so that fluids escape into the surrounding tissues causing swellings such as are seen in allergic reactions, with which it is associated.

HORMONES: Substances secreted by the endocrine glands and acting as chemical messengers in the body.

ICOSAHEDRON: A figure with twenty sides.

JACKEROO: A young man being trained in the pastoral industry in Australia by working on a cattle or sheep station. For young women, the term is ''jillaroo''. These terms though more or less fanciful are accepted and are in common use.

KOCH: (1843–1910) A German country doctor who became one of the world's most successful bacteriologists. He discovered the tubercle and the anthrax bacilli, the cholera vibrio, and the staphylococci, and his methods of culturing, staining and identifying organisms are in constant use.

KOOKABURRA: The native name for a large kingfisher common in Australia and well known for its amusing habit of gathering into groups at sunrise and sunset and laughing loudly in unison. A common local name is ''laughing jackass'', its cackling notes closely resembling a laugh.

LEISHMANIA: Small non-flagellated oval parasites three to five microns in diameter which assume various forms, becoming elongated and flagellated like the trypanosomes with which it shows an evolutionary link.

LEUCOCYTES: (Gk leucos = white; cyt = cell) The white cells of the blood. Normal count 4,000 to 10,000 per cubic millimetre. Various types of cell have different functions (see phagocytosis).

MARROW: The fatty tissue within bone cavities wherein blood cells are manufactured.

MEDIA: The nutrient substances used for the cultivation and growth of micro-organisms. Usually in the form of broths and jellies.

MEDULLA OBLONGATA: The vital brain stem where brain and spinal cord join. The automatic nerve centres as for the control of breathing and heart action, are situated here.

MENINGES: The three membranes which envelop and protect the spinal cord and brain.

MENINGITIS: Inflammation of these membranes.

MEREZOITES: Forms assumed in the asexual reproduction phase of certain protozoa such as the malaria parasites.

MIASMATIC: Pertaining to poisonous vapours, especially those supposed to arise at night from swamps and marshes.

MICRON: The one thousandth part of a millimetre.

MILLIMICRON: The one thousandth part of a micron.

MICROORGANISMS: Any organisms not visible except by the aid of a microscope.

MICROSCOPIC DYES: Dyes, usually anilines, used to make microorganisms more easily seen under the microscope and to assist to identify them by their varied affinities for various dyes.

MITOSIS: The process of cell division in which the chromosomes are split in such a way as to provide similar hereditary characters in each new cell.

MUTATION: Unexpected variation in hereditary characters.

MYOCARDITIS: Inflammation of the heart muscle.

NEUROSYPHILIS: The late stages of syphilis in which the brain and spinal cord are severely damaged.

NIDUS: A nest or breeding place. In bacteriology, may signify a site of infection where germs of disease are harboured.

NUCLEIC ACIDS: The nucleic acids designated D.N.A. and R.N.A. form the molecules which, enveloped in a coat of protein, may constitute the virus particle. D.N.A. is found also in the nuclei of other microorganisms.

OEDEMA: A condition in which for various pathological reasons, body fluids seep out into loose tissues and cause swellings of the affected parts, instead of remaining in circulation.

OXYTOCIN: A hormone secreted by the pituitary gland which enhances uterine contractions.

PANDEMICS: Epidemics which spread quickly around the world. The great influenza pandemic which followed the first world war is a good example.

PERICARDIAL FLUID: The fluid enclosed in the pericardial sac which protects and lubricates the heart action.

PETERLOO: Peterloo or the Manchester Massacre. On August 16th, 1819, five years after Waterloo, a meeting of some 6,000 persons was held in St. Peters Fields, Manchester to seek parliamentary reform. The authorities were seized with panic and ordered the Cheshire yeomanry and the 15th Hussars to charge the crowd. The number of killed and wounded was never known, but it was above 600. The massacre stirred up such indignation that Peterloo was a major factor in finally forcing reforms.

PHAGOCYTES: See leucocytes.

PHAGOCYTOSIS: Literally ''eating cells''. The process by which the leucocytes of the blood actually engulf and digest invading bacteria or other foreign material.

PLEOMORPHIC: Taking on various shapes.

PLUS-FOUR SYPHILIS: The worst grade of an empirical standard for grading syphilis.

PRIMORDIAL CELLS: The earlier cells to appear in the development of a new individual or any of its organs.

PROTEOLYTIC: Capable of breaking down proteins into simpler substances.

PROTOPLASM: The essential semi-fluid colloid in all animal and plant cells.

PUERPERAL: During the period immediately following childbirth.

PURULENT: Discharging pus.

PYREXIAL: Feverish.

PYRRHIC VICTORY: In 281 B.C. Pyrrhus, King of Epirus, defeated the Romans at Heraclea, but with such heavy loss that it was really a defeat and gave rise to the phrase "a Pyrrhic Victory".

RED CELLS: The erythrocytes or oxygen-carrying cells of the blood.

R.F.D.S.: The Royal Flying Doctor Service of Australia which provides a medical service by air to many remote areas of the continent.

RHOMBIC TRIACONTAHEDRON: A geometrical structure having thirty sides in the form of equilateral parallelograms with acute angles.

RICKETTSIA: These organisms, midway in size and characters between viruses and bacteria were once considered to be intermediate forms, but have now been shown to be true bacteria containing both nucleic acids D.N.A. and R.N.A. as in bacteria and not only one as in viruses. The natural reservoirs of rickettsia are in arthropod hosts such as lice, ticks, and fleas, but they infect a large range of animals, and in man produce several groups of distinct and often fatal illnesses, the worst of which is typhus fever.

RINGER: An expert and experienced stockman.

R.N.A. (see also D.N.A.) Ribonucleic acid, one of the two chemical substances of the chromosomes which is responsible for the functioning and reduplication of the genes.

SABIN: The vaccine against poliomyelitis named after SABIN which contains living attenuated poliomyelitis virus to be given by mouth.

SALK: The vaccine against poliomyelitis prepared by SALK and named after him, which contains inactivated poliomyelitis virus for injection. These vaccines have been equally successful in preventing poliomyelitis.

SAPROPHYTES: Organisms which break-up and feed on decaying organic matter.

SEPTICAEMIA: (septicos (Gk) = putrifactive; haemo (L) = blood) Invasion of the blood by pathogenic organisms, and their toxic products. A common term is "blood poisoning".

SEROTYPE: Serotyping, by testing organisms against prepared anti-sera, is used in identifying them. The antibodies in the serum will precipitate or agglutinate the relevant organisms but will have no effect on others.

SPORULATION: The process of forming spores.

SKULL AND CROSSBONES: Syphilis leaves easily recognised marks on the skeletal system and such evidence of syphilis has been found in the bones of Egyptian mummies.

SPIROCHAETES: These are large thin flexible, helical rods, some of which are free-living and non-pathogenic, while others cause such severe diseases as syphilis and leptopirosis, in man.

SPOROZOITES: The spindle-shaped cells which develop in the sexual phase of the life-cycle of the malaria parasite in the mosquito and when injected at the time of its bite into human blood, commence its asexual phase which accompanies the course of clinical malaria in man.

TETRACYCLINES: The tetracyclines are a group of organic drugs used against infectious diseases. Their action is chiefly bacteriostatic by inhibiting the synthesis of proteins.

TRIMESTER: The first, second and third three-monthly periods of gestation are described as the appropriate trimesters.

TOXAEMIA: The condition in which toxins are present in the blood stream.

TOXIGENIC: Capable of producing toxins.

TOXINS: Toxins are various protein substances produced by animals and plants e.g. snake venom, but in bacteriology refer to those poisonous substances manufactured by some microorganisms under suitable conditions. They may be 'exotoxins' which are secreted by the organism into the surrounding tissues such as the exotoxins of diphtheria and tetanus or 'endotoxins' which are contained within the organism and released only when it disintegrates, such as the endotoxins of bacillary dysentery. Exotoxins are many times more potent than endotoxins. The healthy body will synthesise neutralising substances known as 'anti-toxins' if allowed time to do so.

TRACHEOTOMY: The operation of opening the trachea to allow air to the lungs as in cases of choking.

TRANSCEIVER: A wireless instrument for both receiving and transmitting. A piece of standard equipment in the early days of the Royal Flying Doctor Service in Australia for which it was specially designed.

TRANSOVARIANALLY: Transferred from mother to offspring via the ovaries — that is through the egg.

TRYPANOSOMAIASIS: Infection by trypanosomes, the flagellate protozoa which inhabit the blood and other body fluids of man and animals causing fatal illness. Some animals such as antelopes, seem to be unaffected by their presence in the body.

TSETSE FLY: Blood sucking African flies of the genus Glossina which are vectors of trypanosomes.

TRANSPLACENTAL IMMUNITY: Antibodies formed in the body of the mother in response to infection are transferred to the offspring in utero by way of the placental circulation. These protect the offspring for a period after its birth, but like all "passive" types of immunity it is not permanent.

TURKEY NESTS: In the extensive flat plain country in Australia bore water for stock is often stored in elevated earth "tanks" made by building up earth walls above the ground level. Windmills pump water into these "tanks" and it gravitates under valve control into water troughs, so giving an automatic continuous water supply. As the plain turkey builds up its nest in a similar fashion, the appropriate name of turkey nests is applied to these "tanks".

UNDULATING MEMBRANE: The membrane present on some protozoans, such as the trypanosomes, which by its waving undulating action, propels the organism through the fluids in which it lives.

VECTOR: Anything that acts as a living carrier of disease-producing organisms, such as fleas carrying plague bacilli or mosquitoes carrying the virus of yellow fever.

VIBRIO: The vibrios are a large group of curved gram-negative aerobic rods with a single flagellum. Except for vibrio cholera which produces cholera in man, they are only mildly pathogenic but they may affect other animals, such as horses and cattle.

VILLI: Minute finger-like spongy processes for the secretion or the absorption of body fluids, for example, in the intestinal wall for absorption of nutrients and in the placental sinuses for the exchanges between maternal and foetal circulations.

VIRAL CAPSIDS: These are protein shells which encase the nucleic acid core of virus particles. They are extremely complex, being made up of hundreds of structural units arranged in an intricate symmetrical pattern.

VIRAL HEPATITIS: A severe inflammation of the substance of the liver, due to the presence of a specific virus. It may be caused by serum, or be contracted as an infection.

VIROLOGIST: A scientist engaged in the study of viruses.

VIRULENT: Highly infectious and pathogenic. The virulence of any organism is not always constant and can be increased or diminished. Broadly, it may be said that growth under conditions which are artificial for that organism will reduce its virulence, and this is the basis for the production of most immunising vaccines.

VIRUS: The smallest known living units; and probably the smallest capable of independent existence. They consist of a molecule of either R.N.A. or D.N.A. enclosed in a protective coat, and some scientists now think at least some of the smaller viruses may be merely free living chromosomes. They range in size from 10 to 200 millimicrons and therefore may just reach the lowest limit of visibility by the microscope. Viruses can survive and be transferred from host to host in the free state, but can grow and multiply only within the cells of hosts which are susceptible to that particular virus. They cause diseases, probably in every kind of plant or animal. In humans e.g. smallpox, measles, mumps, influenza, poliomyelitis and yellow fever — in plants, tobacco mosaic, leaf roll in potatoes, bunchy top in bananas — in animals, swine fever, foot and mouth disease, distemper in dogs and cats, and many other conditions in living things, such as silkworms and bees.

WARRIGAL: The Australian native dog or dingo.

WAVE LENGTH OF LIGHT: The distance between successive waves. The unit of measurement of the wave-length of light, is the Angstrom unit which is one hundred-millionth of a centimetre.

WHITE CELL COUNT: There are several kinds of white cells or leucocytes in human blood and their functions vary from the destruction of invading foreign matter, such as bacteria, or the repair of damaged tissue, to the manufacture of protective antibodies. Their numbers should be constant at about eight thousand per cubic millimetre and the proportions of the different types should also be constant. Any significant variation of either number or proportion of cells, may indicate abnormal or diseased conditions, so that diagnostic checks on white cells (and also red cells) q.v. are regular requirements in medical practice. The average diameter of white cells varies from six to fifteen microns.

YOLK SAC: The sac which, in the fertilised egg contains the yolk or food for the growing embryo. In the hen's egg, it is an ideal medium for growth of certain organisms, such as rickettsia, which are otherwise difficult to grow, and it is extensively used in research medical laboratories for this purpose.